Atomic Force Microscopy

A New Look at Microbes

Synthesis Lectures on Materials and Optics

Atomic Force Microscopy: A New Look at Microbes

Ahmed Touhami

ISBN: 978-3-031-01257-0   paperback
ISBN: 978-3-031-02385-9   ebook
ISBN: 978-3-031-00249-6   hardcover

DOI 10.1007/978-3-031-02385-9

A Publication in the Springer series
SYNTHESIS LECTURES ON MATERIALS AND OPTICS

Lecture #3
Series ISSN
Synthesis Lectures on Materials and Optics
Print 2691-1930   Electronic 2691-1949

Atomic Force Microscopy

A New Look at Microbes

Ahmed Touhami
University of Texas Rio Grande Valley (UTRGV)

SYNTHESIS LECTURES ON MATERIALS AND OPTICS #3

ABSTRACT

Over the last two decades, Atomic Force Microscopy (AFM) has undoubtedly had a considerable impact in unraveling the structures and dynamics of microbial surfaces with nanometer resolution, and under physiological conditions. Moreover, the continuous innovations in AFM-based modalities as well as the combination of AFM with modern optical techniques, such as confocal fluorescence microscopy or Raman spectroscopy, increased the diversity and volume of data that can be acquired in an experiment. It is evident that these combinations provide new ways to investigate a broad spectrum of microbiological processes at the level of single cells. In this book, I have endeavored to highlight the wealth of AFM-based modalities that have been implemented over the recent years leading to the multiparametric and multifunctional characterization of, specifically, bacterial surfaces. Examples include the real-time imaging of the nanoscale organization of cell walls, the quantification of subcellular chemical heterogeneities, the mapping and functional analysis of individual cell wall constituents, and the probing of the nanomechanical properties of living bacteria. It is expected that in the near future more AFM-based modalities and complementary techniques will be combined into single experiments to address pertinent problems and challenges in microbial research. Such improvements may make it possible to address the dynamic nature of many more microbial cell surfaces and their constituents, including the restructuring of cellular membranes, pores and transporters, signaling of cell membrane receptors, and formation of cell-adhesion complexes. Ultimately, manifold discoveries and engineering possibilities will materialize as multiparametric tools allow systems of increasing complexity to be probed and manipulated.

KEYWORDS

atomic force microscopy, single-molecule force spectroscopy, single-cell force spectroscopy, cell surface nanostructures, molecular recognition, microbial adhesion, cellular interaction, bacterial biofilms, cell mechanics

To my parents, Zohra El Majdoubi & Mohammed Touhami.
To my wife Zohra and my children Rim, Rami, and Rayyan.

Contents

Preface

The nanoscale analysis of microbial cells by atomic force microscopy (AFM) is an exciting, rapidly evolving research field. Compared with other types of microscopy, AFM offers two unique features: (1) the ability to work directly at nanometer resolution in aqueous solutions and (2) the possibility of probing various properties and interactions at the single-molecule level. In the imaging mode, AFM can visualize the surface ultrastructure of live cells under physiological conditions and allows real-time imaging to follow dynamic processes such as cell growth, and division and effects of drugs and chemicals, which opens up new possibilities for studying the assembly and remodeling of cell walls, and for understanding the action mode of antibiotics. AFM is more than a surface-imaging tool in that when used in the force spectroscopy mode, it allows measurement of physicochemical properties of a single cell, such as surface energy and surface charge, mechanical properties, and localization of molecular recognition events. These measurements provide new insight not only in microbiology, to elucidate cellular functions (such as ligand-binding or biofilm formation), but also in medicine (biofilm infections) and biotechnology (cell aggregation).

Recent developments in AFM have been accomplished through various technical and instrumental innovations, including high-resolution and recognition imaging technology under physiological conditions, fast-scanning AFM, and general methods for cantilever modification and force measurement. All these techniques are now highly powerful not only in specific research areas but also in basic biological sciences. There are several AFM books that focus on materials, instruments, and basic biological sciences, but only a few of them are directed toward microbial studies. This book tries to bridge this gap. Writing by a leading expert in the field, this book provides an overview of modern AFM technologies for microbial cell analysis, going from the basic principles to the applications. The different chapters cover the most recent methodologies for preparing and analyzing microbial cells, discuss the principles of advanced AFM modalities, including high-resolution imaging, high-speed imaging, recognition imaging, cell-cell adhesion, mechanical measurements, and highlight recent applications in a variety of fields, including cell biology, microbiology, biophysics, structural biology, physiology, and medicine.

I hope that the book will interest students and researchers from various horizons, whether they are newcomers or well trained in the field. The volume should help them to evaluate the advantages and limitations of AFM techniques in their specific field and to define appropriate procedures and controls that will lead them to successful experiments.

Ahmed Touhami
May 2020

Acknowledgments

First, I would like to thank God for his mercy and blessing on me to complete this piece of work. I would like to extend my warmest thanks to all my teachers, mentors, and advisors, who sacrificed their time and energy to guide me in this long learning journey. I would also like to express my heartfelt gratitude to my parents and all family members—thank you for your love, support and prayers. Finally, I thank my graduate students who spent time helping to finish the book.

Ahmed Touhami
May 2020

C H A P T E R   1

Measurement Methods in Atomic Force Microscopy

OVERVIEW

Optical microscopes are an iconic piece of equipment in any laboratory and are one of the most influential scientific inventions ever made. Almost everyone can relate to the excitement of the first time they looked into the lens of a microscope and experienced for the first time the wonders of a hidden world not otherwise visible to the human eye. However, compared to the microscopes developed in the 19th century, the conventional optical microscopes used to date have no substantial difference or improvement, because the optical microscopes have reached their maximum limitation in spatial resolution. The invention of microscopes not only extended our sense of seeing but also revolutionized our perception of the world. Extending this perception further and further has since been the driving force for major scientific developments. The dawn of the 20th century witnessed the development of an alternative to the light microscope known as the electron microscope. These devices used electrons rather than light to generate an image of the target. The development of this technology was representative of a huge advance in the field, given that these microscopes were capable to delivering a much higher resolution. Because electrons can be accelerated to very high speed, the spatial resolution of electron microscopes reaches up to 0.3 nm. As a result, many invisible materials under the visible light become "visible" under the electron microscope.

In the 1980s, local probe microscopy techniques extend our sense of seeing to touching the micro- and nano-world and in this way provide complementary new insight into these worlds with microscopic techniques. Furthermore, touching things is an essential prerequisite to manipulating things, and the ability to feel and to manipulate single molecules and atoms certainly marks another of these revolutionizing steps in our relation to the world we live in. The first of these local probe instruments was the scanning tunneling microscope (STM), which was invented by Gerd Binning and Heinrich Rohrer in 1983 [1]. This advanced microscope was built using totally different concepts than those used by the conventional microscopes. STM works based on the so-called "tunnel effect." STM has no lens; rather, it uses a probe. Once the voltage is added between the probe and the observed object, the tunnel effect occurs if the distance between the probe and the surface of the observed objects is sufficiently small (few nanometers). When the electrons pass through the tiny space between the probe and the object, weak electronic current is generated. If the distance between the probe and the object varies, the strength of the current varies as well. Hence, the three-dimensional shape of the object can be detected as one measures the electronic current change. The STM for the first time showed the atomic structure at the crystalline surface of silicon in real space and demonstrated that it was even possible to manipulate single atoms. The importance of this development was recognized when the Nobel Prize in Physics was awarded to Binnig and Rohrer in 1986. In the same year, Binnig together with Quate and Gerber demonstrated that the short-range van der Waals interaction can also be used to build a scanning probe microscope [2]. This new device was called the atomic force microscope (AFM). With no electron transport involved, even insulators could be studied down to atomic resolution. AFM has revolutionized surface characterization by allowing the researcher to examine the molecular structure of virtually any sample under virtually any environmental condition. The AFM is used to produce information about surface topography, elasticity, friction, adhesion, charge density, magnetic structure, or even long-range effects. AFM has thus undergone several developmental stages, and due to its relative simplicity, it is a fundamentally versatile technique, with it being difficult to foresee the limits of its evolution and application to different areas of knowledge.

This chapter will introduce the main principles and operation modes of the atomic force microscopy and review the progress over the last decades in the development and the integration of AFM into hybrid devices.

1.1   PRINCIPLES AND MODES OF OPERATION

As shown in Figure 1.1, the essential part of an AFM, as for all scanning probe microscopes, is the tip that determines by its structure the type of interaction with a sample, and by its geometry, the area of interaction. The sample is scanned by a tip, which is mounted to a cantilever spring. While scanning, the force between the tip and the sample is measured by monitoring the deflection of the cantilever. A topographic image of the sample is obtained by plotting the deflection of the cantilever vs. its position on the sample. Alternatively, it is possible to plot the height position of the translation stage. This height is controlled by a feedback loop, which maintains a constant force between tip and sample. Image contrast arises because the force between the tip and sample is a function of both tip-sample separation and the material properties of tip and sample. To date, in most applications image contrast is obtained from the very short-range repulsion, which occurs when the electron orbitals of tip and sample overlap. However, further interactions between tip and sample can be used to investigate properties of the sample, the tip, or the medium in between. These measurements are usually known as "force measurements" [3], and will be detailed later in this chapter.

A similarly important part of the scanning probe microscope is the piezo-tube scanner used to produce movements in all three directions easily and consists of a thin-walled hard piezo-electric ceramic that is radially polarized. The piezo moves the tip closer to the surface and scans it across with precision fitting to the highest resolution.

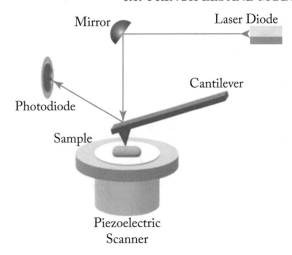

Figure 1.1: Schematic of an atomic force microscope.

Many types of scanning probe microscopes have been developed and can be used not only for measuring surface topologies but also for measuring various material properties at or close to surfaces [4]. This can be done in vacuum, gas, or liquids in a broad temperature range with a resolution down to either the atomic or the molecular level. In this way, it is the only type of microscopy that can complement optical microscopy in biology on a smaller scale. Additionally, these instruments allow manipulations at either the single-atomic or the molecular level, making experiments which no one ever dreamed of 30 years ago possible. Experiments at the nanometer scale provide completely new insight into processes which, before the development of these instruments, were accessible only by ensemble-average processes, where all of the elements can never be identical, and all of the information concerning the behavior of individuals is lost. With the available information on single components using scanning probe techniques we can now learn how processes, which we were previously unaware of, are determined by the properties of the single elements of such ensembles [5].

One can easily distinguish between two general modes of operation of the AFM depending on absence or presence in the instrumentation of an additional device that forces the cantilever to oscillate in the proximity of its resonant frequency. The first case is usually called static mode, or DC mode, because it records the static deflection of the cantilever, whereas the second takes a variety of names among which we may point out the resonant or AC mode. In this case, the feedback loop will try to keep at a set value not the deflection but the amplitude of the oscillation of the cantilever while scanning the surface. To do this, additional electronics are necessary in the detection circuit, such as a lock-in or a phase-locked loop amplifier, and also in the cantilever holder to induce the oscillatory excitation. From a physical point of view, one can make a distinction between the two modes depending on the sign of the forces involved in the

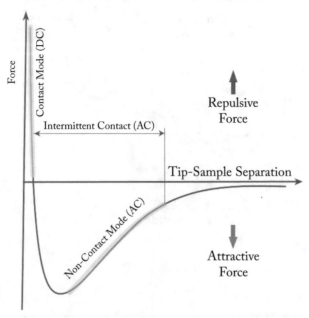

Figure 1.2: An idealized plot of the forces between tip and sample, highlighting where typical main imaging modes are operative.

interaction between tip and sample, that is, by whether the forces there are attractive or repulsive. In Figure 1.2, an idealized plot of the forces between tip and sample is shown, highlighting where typical imaging modes operate. In the following we briefly describe the DC, AC, and other modes of operation.

1.1.1　CONTACT MODE

Also called DC mode, the contact mode is the original and simplest mode to operate an AFM. In this mode, the tip is in continuous contact with the sample while the tip raster scans the surface (Figure 1.2). The most common configuration is to operate in constant force or deflection feedback mode where the cantilever deflection is kept constant during scanning by the feedback loop. The cantilever deflection is set by the user and is related to how hard the tip pushes against the surface so that the user controls how gentle or aggressive the interaction between the probe and the sample. Image contrast depends on the applied force, which again depends on the cantilever spring constant. Softer cantilevers are used for softer samples.

　　Contact mode can also be operated in constant height mode where the probe maintains a fixed height above the sample. There is no force feedback in this mode. Constant height mode is typically used in atomic resolution AFM, though it is uncommon for other AFM applications.

Finally, there is a configuration known as "error mode." In some cases, especially on rough and relatively rigid samples, the error signal (i.e., the difference between the set point and the effective deflection of the cantilever that occurs during scanning as a result of the finite time response of the feedback loop) is used to record images. By turning down on purpose the feedback gain, the cantilever will press harder on asperities and less on depressions, giving rise to images that contain high-frequency information otherwise not visible. This method has been extensively used to image submembrane features in living cells.

The contact mode can be used easily also in liquids, allowing a considerable reduction of capillary forces between tip and sample and, hence, damage to the surface. Because the tip is permanently in contact with the surface while scanning, a considerable shear force can be generated, causing damage to the sample, especially on very soft specimens' like biomolecules or living cells.

A variation of standard contact mode is the lateral force microscopy, in addition to the vertical deflection of the cantilever, the lateral deflection (torsion) is measured by the lateral photodetector assembly. The degree of torsion of the cantilever supporting the probe is a relative measure of surface friction caused by the lateral force exerted on the scanning probe. This method has been used to discriminate between areas of the sample that have the same height (i.e., that are on a same plane) but that present different frictional properties because of absorbates [6].

1.1.2   DYNAMIC MODE

Shortly after the invention of the AFM, efforts were directed at overcoming the drawbacks associated with contact mode operation and increasing the amount of physical information that can be extracted from experiments. This led to the introduction of dynamic modes in which the cantilever oscillates at a high frequency at or close to resonance. The most common dynamic mode is the so-called tapping mode (TM-AFM) in which the amplitude of oscillation is the feedback parameter. TM-AFM is also referred to as AC mode or intermittent contact mode. Other dynamic modes have different parameters for feedback such as frequency (frequency modulation) or phase (phase modulation) [4].

As an imaging mode, tapping mode offers several key advantages. Because the cantilever operates at resonance and interacts with the sample as the probe "taps" along the surface, it is a gentle interaction with the surface relative to static imaging modes that can preserve the sharpness of the tip. This kind of interaction also minimizes torsional forces between the probe and the sample, which are especially exacerbated in static imaging mode. These two advantages are particularly important for soft materials such as bacteria cells, biopolymers, or fibrillar samples where dynamic modes are less destructive to the sample. Finally, by using the cantilever's oscillation amplitude as the feedback parameter, the user is able to fine-tune the interaction between probe and sample between different regimes such as attractive and repulsive regimes.

Phase Imaging Mode

If the phase lag of the cantilever oscillation relative to driving signal is recorded in a second acquisition channel during imaging in intermittent contact mode, noteworthy information on local properties, such as stiffness, viscosity, and adhesion, can be detected that are not revealed by other AFM techniques. In fact, it is good practice to always acquire simultaneously both the amplitude and phase signals during intermittent contact operation, as the physical information is entwined, and all the data is necessary to interpret the images obtained.

Noncontact Mode

A fundamental shortcoming of both contact and tapping mode AFM involves the formation of a finite contact area between the tip apex and the sample surface during operation. Thus, a tip that is atomically sharp at the onset of the experiment becomes blunt during operation, resulting in the inability to achieve atomic resolution via contact and tapping modes in most cases. To achieve atomic-resolution imaging, the dynamic operation mode of "noncontact" AFM (NC-AFM) has been introduced [4]. In this mode, an oscillating probe is brought into proximity of (but without touching) the surface of the sample and senses the van der Waals attractive forces that induce a frequency shift in the resonant frequency of a stiff cantilever. Images are taken by keeping a constant frequency shift during scanning, and usually this is performed by monitoring the amplitude of the cantilever oscillation at a fixed frequency and feeding the corresponding value to the feedback loop exactly as for the DC modes. The tip-sample interactions are very small in noncontact mode, and good vertical resolution can be achieved, whereas lateral resolution is lower than in other operating modes.

Force Modulation

In this case, a low-frequency oscillation is induced (usually to the sample) and the corresponding cantilever deflection recorded while the tip is kept in contact with the sample. The varying stiffness of surface features will induce a corresponding dampening of the cantilever oscillation, so that local relative viscoelastic properties can be imaged.

1.2   FORCE MEASUREMENT AND FORCE MAPPING

The AFM is not only a tool to image the topography of surfaces at high resolution and under various conditions. It can also be used to measure forces between different surfaces at the nanoscale accurately down to few pN. Such measurements, briefly called force spectroscopy, provide valuable information on local material properties such as inter-molecular forces, adhesion, elasticity, and surface charge densities [7, 8]. Force spectroscopy can also be used to probe the local chemical groups, and receptor sites of live cells. Other applications locate molecular interactions driving membrane protein folding, assembly, and their switching between functional states. It is also possible to examine the energy landscape of biomolecular reactions, as well as reaction pathways, associated lifetimes, and free energy. There are several features of AFM that make it ideal for force sensing, including: sensitivity of the displacement of ca. 0.01 nm, small tip-sample contact area of about 10 nm², and the ability to operate under physiological conditions. For these reasons, AFM force spectroscopy has become essential in different fields of research such as biology, biophysics, surface science, and materials engineering.

In order to analyze AFM force experiments, one has to understand how single-point force-distance curves are obtained, and the information provided regarding tip-sample interaction. In this paragraph we describe the basics of AFM force measurements. Current state of the art in analyzing force curves obtained under different conditions will be presented in next chapters.

1.2.1   FORCE-DISTANCE CURVE

In AFM-based force measurements the sample is moved up and down by applying a voltage to the piezoelectric translator, onto which the sample is mounted, while measuring the cantilever deflection. In some AFMs the chip to which the cantilever is attached is moved by the piezo-electric translator rather than the sample. The design of a force-distance curve depends on the tip, sample, and medium composition. The curves obtained for samples in air vary from those obtained in liquid medium. In air, as the cantilever approaches the surface, the initial forces are too small to achieve a deflection and therefore it remains in the same position on the ordinate axis (Figure 1.3). As the sample moves toward the tip various attractive forces pull on the tip (long- and short-range forces). Once the total force acting on the tip exceeds the stiffness of the cantilever it jumps into contact with the sample surface (jump-to contact) (Figure 1.3, B–C). At point (D), the tip and sample are in contact and deflections are dominated by mutual electronic repulsions between overlapping molecular orbitals of the tip and sample atoms (A–D is the approach curve). In this region, which is typically linear, the elastic properties of the sample can be measured.

During withdrawal, adhesion or bonds formed during contact with the surface cause the tip to adhere to the sample up to some distance beyond the initial contact point on the approach curve (E–F). As the piezotube continues retracting, the spring force of the bent cantilever overcomes the adhesion forces, and the cantilever pulls off sharply, springing upward to its non-deflected or noncontact position (G). Finally, the tip completely loses contact with the surface and returns to its starting equilibrium position (H) (D–H is the withdrawal curve).

The result of a force measurement is a measure of the cantilever deflection, Z_c, vs. position of the piezo, Z_p, normal to the surface. To obtain a force-versus-distance curve, Z_c and Z_p have to be converted into force and distance. The force F is obtained by multiplying the deflection of the cantilever with its spring constant k_c (Equation (1.1)). The tip-sample separation distance D is calculated by adding the deflection to the position as in Equation (1.2):

$$F = k_c \times Z_c \tag{1.1}$$

$$D = Z_p + Z_c. \tag{1.2}$$

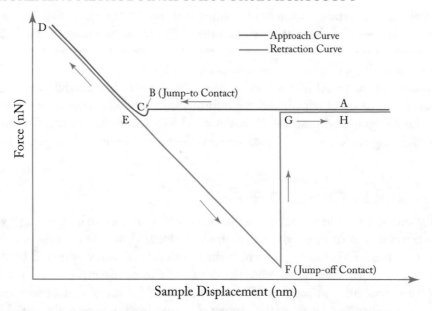

Figure 1.3: Example of force-displacement curve in air illustrating the points where jump-to-contact (approach) and jump-off-contact (withdrawal) occur and the maximum values of the attractive forces (pull-on force and pull-off force).

To acquire force curves in liquid different types of liquid cells are employed. Typically, liquid cells consist of a special cantilever holder and an O-ring sealing the cell. The curves in liquid medium follow the same principle as described above for the air curves, although the approach curve has a different curvature. It presents a gradual increase in strength and, thus, it is difficult to establish the point where the tip and sample come into contact, as the initial compression of the surface causes a minimum deflection in the cantilever. When the tip is removed from the surface by the repulsive forces, there is a delay on the response of the system due to the effect of the elasticity of the sample or of the long-range forces, and different force curve responses can be observed.

Although AFM provides an absolute measurement of the tip position (x, y, z), it is often a challenge to determine the exact contact point between tip and sample (zero separation), particularly when long-range surface forces, surface roughness, and deformation of the soft biological sample play roles. Knowledge of the contact point is needed to differentiate surface forces from the mechanical deformation of the soft cell. However, for most applications, linearly extrapolating the contact region to zero force is sufficiently accurate.

Forces on the Approach Curve

van der Waals interactions are the main type of force present on the approaching of two hard surfaces in the absence of long-range interactions [3]. This force is characterized by a small deflection of the cantilever (jump to contact), at the approach curve, before the contact point. If the approach curve has a smooth and exponentially increasing repulsive force, it is expectable that electrostatic or polymer-brush forces are present [3]. With AFM force curves, some material properties of samples can be investigated by linking the applied force to the depth of indentation as the tip is pressed against the sample. Given a calibrated sensor response, the shape of segment (C–D) indicates whether the sample is deforming in response to the force from the cantilever. This slope can be used to infer the sample hardness or indicate a differing sample response at different loading charges, which result from a transition from elastic to plastic deformation. The segment (D–E) represents the reverse of segment (C–D). If both segments are straight and parallel to each other, there is no additional information content. If they are not parallel, the hysteresis gives information about the plastic deformation of the sample [7]. Recently, AFM approach force curves were intensively used to determine the viscoelastic properties of many biological systems, including cartilage, gelatin gels, glial cells, and epithelial cells [9, 10].

Forces on the Retraction Curve

It is common to see on the retraction curve an adhesion force profile due to hysteresis. This force depends on the type of the sample and appears as a deflection of the cantilever below the zero-deflection line. The central basis of adhesion forces is the development of a capillary bridge between the tip and the sample. This capillary force depends if the measurements are made in air or in liquid. In air, samples usually have several nanometers of water molecules adsorbed to their surface. This water layer forms the bridge between the tip and the sample, with a large force being required to overcome the surface tension and pull out the tip from the sample [7]. In liquid conditions, the adhesion force not only depends of the interaction energies between tip and sample but also on the type of solution used. When a polymer is present connecting the tip and the surface, a negative deflection of the tip at a given distance away from the sample is obtained. The difference in distance of the observed deflection is due to the polymer extension. The polymer is stretched until the bond breaks or the polymer detaches from the tip or the surface. After that, the cantilever returns to the zero-deflection line position. Examples of all these types of force curves are detailed at [7]. To quantitatively analyze the force-curves, some specific requirements are necessary, such as the accurate calibration of the spring constant of the cantilevers used. Cantilever stiffness depends on the shape and on the material properties of the cantilevers. Although for commercial cantilevers, a specific spring constant value is given on purchase, different values between cantilevers from the same batch can be experimentally observed. Thus, to properly measure the force values associated with a force-distance curve, it is necessary to use a specific spring constant calibration. Also, if the tip is coated by functionalization processes, the cantilever spring constant can be modified. The

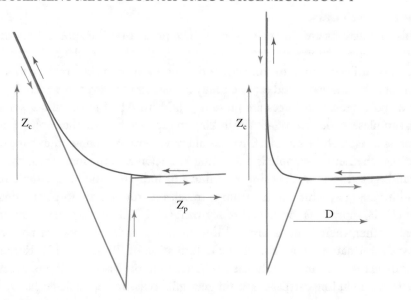

Figure 1.4: (a) Schematic of a typical cantilever deflection vs. piezo height (Z_c-vs.-Z_p) curve; and (b) Corresponding cantilever deflection vs. separation distance D curve.

most commonly used calibration method is the thermal noise method, that can be used in air or in liquid, with the precision of the measurements varying between 10% and 20%.

1.2.2   FORCE VOLUME MAPPING

Maps of the distribution of intra- and intermolecular forces can be collected over the sample scanning area by force volume imaging mode. Individual force curves are obtained at unique x–y positions on the sample, so that interaction forces can be measured at each location. This array of force curves forms a three-dimensional force map illustrating the distribution of any particular interaction of interest. For example, most simply an array of adhesion forces extracted from force-distance curves would produce an adhesion map of the sample. Through selective modification of the AFM probe, this adhesion map can be tailored to isolate a specific interaction thereby showing the distribution of an otherwise unknown entity. Figure 1.4 illustrates a schematic representation for adhesion mapping of proteins in a multi-component protein layer on a material surface. In this case, the probe is modified with an antibody of interest, and interactions between the antibody-modified probe and the substrate demonstrate the presence of the desired antigen.

1.3 AFM-NANOMECHANICAL MAPPING

As mentioned in the previous section, AFM can be used to measure elasticity of surfaces. Compared with other tools, AFM can probe local surface mechanical properties with high resolution, down to several nanometers, and with fine control of applied force. These two characteristics give the AFM advantages for studying the mechanical properties of polymeric and biological systems because most of these exhibit nanoscales heterogeneous modulus distribution.

1.3.1 NANOMECHANICAL MEASUREMENT PROCEDURE

The easiest way to measure mechanical properties by AFM is to indent the probe into the sample and to record the applied force, which is proportional to the cantilever deflection, and the distance traveled by the probe in a force-distance (FD) curve (Figure 1.5a,b). Recorded upon approaching and retracting the probe, FD curves measure the mechanical deformation and response of the sample under load. Force can also be plotted against time in force-time (FT) curves, which are particularly useful if the force applied by the indenting probe or the indentation depth of the probe is to be held constant [10, 11] (Figure 1.5c,d). These mechanical readouts are particularly useful when the sample changes mechanical properties with time [12, 13] or viscoelastic properties need to be determined [14, 15]. The indentation depth (δ) can be calculated from the z-piezo movement (Z_p) and the cantilever deflection (Z_c) as:

$$\delta = Z_p - Z_c. \tag{1.3}$$

By analyzing the depth-sensing indentation and force spectroscopy, the local elasticity of the sample in terms of the Young's modulus can be obtained from various classic models including the Hertz, Sneddon's, Johnson–Kendall–Roberts (JKR), and Derjaguin–Muller–Topov (DMT) models [16, 17]. The simplest of these models is the Hertz model which describes a sphere indenting an infinite flat surface with a purely elastic interaction. Derivation of Young's modulus from the Hertz model gives [16]:

$$E = \frac{3}{4}\left(1 - \mu^2\right)\frac{k}{R^{\frac{1}{2}}}\frac{Z_{defl}}{\delta^{\frac{3}{2}}}, \tag{1.4}$$

where (μ) is Poisson's ratio, k is the spring constant of cantilever, and R is the tip radius. The Hertz model is valid for the elastic deformation of the sample.

The Sneddon's model gives the relationship between indentation load gradient, $dF/d\delta$, and Young's modulus, E, in the form [18]:

$$\frac{dF}{d\delta} = 2\frac{\sqrt{A}}{\sqrt{\pi E}}, \tag{1.5}$$

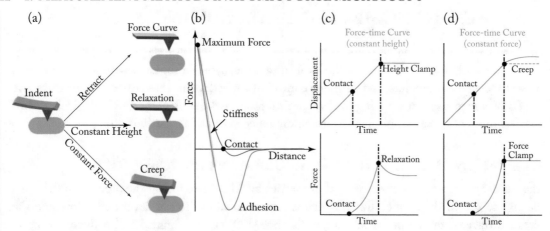

Figure 1.5: (a) Schematic illustration of AFM-based mechanical measurements. (b) Example of force-distance curve showing approach (blue) and retraction (red) of the AFM tip, several mechanical properties of the sample can be deducted from the force curve such as stiffens and adhesion. (c) Example of time-dependent indentation curve (constant height): the probe indents the sample and is then kept at a constant height. The force recorded by the cantilever quantifies the mechanical response of the sample. (d) Example of time-dependent indentation curve (force-constant): the probe indents (or confines) the sample, and the cantilever is kept at a constant deflection (force). The displacement of the cantilever quantifies the mechanical response of the sample.

where F is the indentation load, A is contact area, and δ is indentation depth. E is the composite elastic modulus and can be defined:

$$\frac{1}{E} = \frac{1 - \mu_{tip}^2}{E_{tip}} + \frac{1 - \mu_{sample}^2}{E_{sample}},$$ (1.6)

where E_{tip}, E_{sample}, μ_{tip}, and μ_{sample} are the Young's moduli and Poisson's ratio of the AFM tip and the studied material sample, respectively.

By estimating $dF/d\delta$ and contact area for specific shape of the indenter (circular, pyramidal, and parabolic), the elastic modulus of material can be evaluated from Equation (1.4). For a polymeric biomaterial and biological sample systems, we can assume $E_{tip} >> E_{sample}$, and therefore, $E = E_{sample}$. Chizhik and his colleagues [16] derived models for pyramidal and parabolic shape of indenters based on Sneddon's model:

$$\textbf{\textit{Pyramidal Indenter}} \quad E = \frac{\sqrt{\pi}}{2\sqrt{2}\beta \tan\alpha}\left(1-\mu^2\right)k\frac{\Delta Z_{defl,i,i-1}}{\delta\Delta\delta_{i,i-1}} \tag{1.7}$$

$$\textbf{\textit{Parabolic Indenter}} \quad E = \frac{\sqrt{\pi}}{2\sqrt{R}\beta}\left(1-\mu^2\right)k\frac{\Delta Z_{defl,i,i-1}}{\sqrt{\delta}\Delta\delta_{i,i-1}} \tag{1.8}$$

$$\textbf{\textit{Conical Indenter}} \quad E = \frac{\left(1-\mu^2\right)\pi k Z_{defl}}{2\tan\alpha\delta^2}, \tag{1.9}$$

where $\beta = A_{cros}/A$, A_{cros} is the cross-sectional area of the indenter at the indentation depth δ from the apex, β is equal 2 for elastic deformation of a sphere and $(\pi/2)^2$ for pyramidal and conic shapes, α is half of the pyramidal angle of the indenter, and i, $i-1$ refers to the adjacent indenter displacements.

The Hertz and Sneddon's models are relatively simple with the assumption of no adhesion or friction, and do not require measurements of interfacial energies. The models are often selected for calculation of elasticity of polymeric material and cell surfaces [19]. However, when adhesion is present between indenter and material, the Hertz and Sneddon's models should be modified to accurate the modulus. The JKR, DMT, and Maugis–Dugdale (MD) models are used to consider the influence of adhesion contact on modulus measurement. Physically, the JKR theory [20] account for adhesion forces within the expanded area of contact whereas the DMT theory [21] accounts for adhesion forces act in an annular zone around the contact but without deforming the profile. Chizhik et al. [16] provided the modulus calculation from derivation of JKR theory:

$$E = \frac{9}{4}\left(1-\mu^2\right)Rk\Delta\left[\frac{P_1}{3R\delta}\right]^{3/2}, \tag{1.10}$$

where $P_1 = (3P_2 - 1)[1/9(P_2 + 1)]^{1/3}$, $P_2 = \left(Z_{defl}/\Delta + 1\right)^{1/2}$, and Δ is the cantilever deflection at the point where the probe loses contact with the surface.

The derivation of DMT theory gives [22]:

$$F = \frac{4}{3}E\frac{a^3}{R} - 2\pi R\Delta\gamma, \tag{1.11}$$

where E is the composite elastic modulus defined from Equation (1.6), F is the indentation load, a is contact radius, R is the sphere indenter radius, and $\Delta\gamma$ is the work of adhesion.

The above nanomechanical models have their own assumptions and limitations in applications, which have been discussed elsewhere [22]. In addition, it should be noted that AFM nanomechanical measurement of thin films or cells is often operated on a stiff substrate used to support the thin films or to culture cells, resulting in a "bottom effect" artifact. To correct such errors, Gavara and Chadwick [22] made the effort to correct the Sneddon's model with bottom effect cone correction and demonstrated a significant improvement in the estimation of the local mechanical properties of cells. In addition, the absolute determination of Young's modulus of all biological system, which is complex, is not always necessary. Very often only a comparison between two different states are needed. In this case, individual force curves can be interpreted into arrays of relative elastic values, which can be related to a topographic image. This kind of "elastic mapping" can provide valuable insight into the biological importance of, e.g., cellular mechanics and their regulation.

1.4   AFM-BASED ELECTRICAL METHODS

The enormous adaptability of AFM microscopes enabled the development of accessories and software dedicated to probe a wide variety of electrical properties of materials and surfaces. Probing properties such as current, conductance, surface potential, and capacitance are increasingly important in a number of applications including research on solar and battery cells, conductive polymers, and biological molecules. AFM electrical methods operate either in static mode or dynamic mode, depending on the information being sought. This also will dictate the form of modulation applied to the AFM cantilever or sample (mechanical, electric, or magnetic modulation). Moreover, operating regimes may be implemented in different configurations depending on the property to be measured (one-terminal, two-terminal, or three-terminal) (Figure 1.6). Thus, the extensive variety of combinations enables the development of several AFM electrical techniques. Electrostatic force microscopy (EFM) and scanning surface potential microscopy (SSPM) are examples of modes based on the interaction of electrostatic forces available for the AFM microscope [23]. One can also include the development of powerful tools to provide additional information on some surface electrical parameters, such as conductive atomic force microscopy (c-AFM), which provide high-resolution local measurements of resistance and conductivity, aiding in the detection of defects in integrated circuits, for example [24].

Thus, there is a vast variety of techniques for electrical measurements using the atomic force microscope, with various sensitivities and spatial resolutions. However, the main limitations of these techniques arise from the complicated geometry of the probe-surface system [25]. The situation is more complicated with biological system that need to be investigate in aqueous conditions.

1.5   HIGH-SPEED ATOMIC FORCE MICROSCOPY

One of the limitations that makes AFM unsuitable for studying many dynamic processes in cell biology is the long image acquisition time of several minutes for one high-resolution AFM image. Despite the enormous success of AFM to produce high-resolution images of live cells and cell fragments, most users want higher speed imaging. There are many experiments, such as watching biological processes in liquids, that simply cannot be done without faster imaging. The productivity and use of AFMs would increase dramatically if the speed could match the millisecond to minute image times of other scanning microscopes such as confocal microscopy.

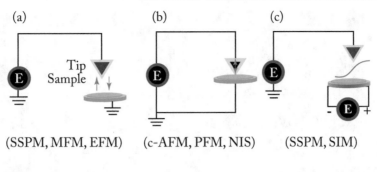

Figure 1.6: **AFM** configurations for electrical modes: (a) one-terminal, (b) two-terminal, and (c) three-terminal. c-AFM, conductive atomic force microscopy; EFM, electrostatic force microscopy; NIS, nano impedance spectroscopy; PFM, piezoresponse force microscopy; SIM, scanning impedance microscopy; SSPM, scanning surface potential microscopy.

The AFM speed is primarily limited by the resonance frequency of the cantilever which is in turn dictated by the cantilever dimensions. Following the relation (for rectangular cantilevers only):

$$f_0 = \frac{1}{2\pi} \sqrt{\frac{k}{m}} \tag{1.12}$$

and

$$k = \frac{E \cdot t^3 \cdot w}{4L^3}, \tag{1.13}$$

where k is the spring constant, m the mass of the cantilever, and t, w, and L, the thickness, width, and length of the cantilever and E the Young's modulus of the cantilever material [26].

Most commercially available AFMs are built to accommodate 50–500 μm long cantilevers, and therefore bio-imaging at nanometric resolution is generally achieved within a timescale of one to several minutes limiting bio-AFM imaging to structural applications of stable samples. The speed limitations of AFM was possible to be overcome using cantilevers of the smallest possible size, yet with soft enough spring constant, high-resolution scanning of biological samples at video rate, termed high-speed atomic force microscopy (HS-AFM), in physiological conditions is now possible [27]. HS-AFM has provided unprecedented insights into the dynamics of membrane proteins and molecular machines from the single-molecule to the cellular level [27] (Figure 1.7). HS-AFM imaging at nanometer-resolution and sub-second frame rate is opening novel research fields depicting dynamic events at the single bio-molecule level. As such, HS-AFM is complementary to other structural and cellular biology techniques and is gaining acceptance from researchers from various fields.

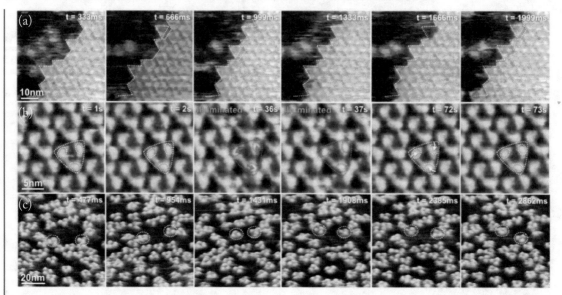

Figure 1.7: HS-AFM imaging of native isolated or reconstituted membranes. (a) Association and dissociation dynamics of bR at purple membrane lattice edges, resulting in a −1.5 kBT interaction energy between molecules. (b) Light-induced conformational changes of bR, revealing movement of the E–F loop with cooperativity of the monomers of adjacent trimers. (c) Lateral and rotational diffusion dynamics analysis of OmpF. Molecular motion scales inversely with the accessible free membrane space. (Reprinted from [27] with permission from Wiley.)

1.6   HYBRIDS OF AFM WITH OTHER TECHNIQUES

AFM has evolved from the originally morphological imaging technique to a powerful and multifunctional technique for manipulating and detecting the interactions between molecules at nanometer resolution. However, AFM cannot provide the precise information of synchronized molecular groups and has many shortcomings in addressing the dynamics of biomolecules due to the limitations of the technology. To overcome these problems, it is necessary to integrate the AFM into hybrid devices utilizing a combination of two or three complementary techniques in one instrument capable to investigate the details of the interactions among molecules and molecular dynamics [28].

Optical microscopy techniques are the most easiest and comment to be integrated with AFM [29]. Modification of the AFM setup to be coupled to an inverted or an upright microscope requires the creation of an obstacle-free optical path to the AFM cantilever and the substrate to be analyzed (Figure 1.8). Tip-scannable atomic force microscopes are preferable when simultaneous optical and AFM imaging is required. In the case of inverted microscopes, a considerable effort in AFM design is needed so as to accommodate optical condensers of

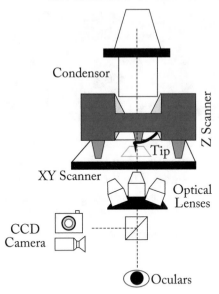

Figure 1.8: Schematic of the combination of inverted optical microscope with an AFM. To ensure simultaneous optical and AFM imaging, the AFM tip can move in three dimensions on XY-stage.

high numerical aperture without compromising the quality of optical imaging or instrument stability. Coupled with AFM's ability to measure high-resolution topographical images, forces, and/or elasticity on a sample, a more complete understanding of structure function relationships can be elucidated with a combined system. While the two imaging modalities have been used in combinational studies for over a decade, significant challenges of direct correlation of the two data sets have existed primarily due to the scaling differences between the two data sets. Recent developments in software now allow for user-friendly and intuitive routines for direct overlay and comparison between the two data sets. Further, various optical techniques are now being used to modify or stimulate samples of interest in concert with AFM measurements, and vice versa [29]. Using transparent substrates, AFM imaging has been combined with the family of transmitted light optical microscopies such as phase and differential interference contrast (DIC), fluorescence, confocal laser scanning microscopy (CLSM), total internal reflection fluorescence (TIRF), and fluorescence lifetime imaging microscopy (FLIM) [29–32]. The use of AFM in combination with optical microscopies and spectroscopic techniques was reviewed in details [33]. Indeed, AFM researches find themselves in a diverse, multi-interfacial area of microscopy, made even more powerful by combining AFM with optical microscopy.

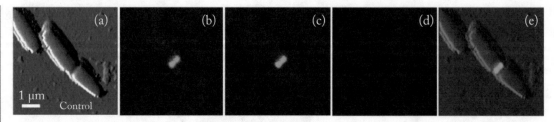

Figure 1.9: Simultaneous AFM-QI-LSCM of E. coli showing the localization of the Z-ring and onset of ROS during 2.4-D exposure. (A-E) Show the presence of a distinct Z-ring (C) in control cells and absence of a ROS signal (D). The confocal image was collected 5 min following the addition of 1 mM 2, 4-D directly into the imaging media. (E) An AFM image overlaid with a confocal one. FtsZ-GFP is shown in green and ROS. Labeled with Cell-ROX, is shown in red. G and B show localization of two colors simultaneously. Bar is 1 μm. (From [36].)

1.6.1   INTEGRATED AFM/CONFOCAL

CLSM has been widely used in biomedical research and life sciences to study and analyze cell morphology, cell components, and cell adhesions and dynamics [34]. Combining confocal and AFM on fluorescently labeled samples has become a unique tool to perform optical and topographic measurements with single molecule sensitivity throughout the whole visible to near-infrared spectral region, and provides the potential precisely link specific chemical or biological information to corresponding topographical features, molecular forces, and mechanical properties, in real time, and under physiological conditions [35]. CLSM makes use of coherent laser illumination and an optical pinhole system to extract thin optical sections without the need for physical sectioning of the specimen. As a result, it is possible to reconstruct a three-dimensional image of the specimen by stacking several of those images taken at different focal planes. Integrated AFM-CLSM is becoming powerful tool for investigating the three-dimensional structures of cellular membranes, from the nanometer to the single-cell levels, as well as for imaging cytoskeleton structures and intracellular features under the cellular membrane. For example, Supriya et al. demonstrated interesting correlative AFM-CLSM setup for the acquisition of nm-resolution surface ultrastructure, pN-scale nanomechanics, and simultaneous localization of two and three intracellular signals in actively dividing bacterial, fungal, and human cells in real time [36] (Figure 1.9). Also, biochemically modified AFM tips make it possible to map or stimulate single molecules on the cell surface while viewing the intracellular response by CLSM, ideal for studying cell signaling across the cell envelope and further multiplexing of the data [37].

1.6.2   INTEGRATED AFM/TOTAL INTERNAL REFLECTION FLUORESCENCE

Total internal reflection fluorescence (TIRF) profits from the evanescent wave generated at a surface by a totally reflected light source. This evanescent wave extends a few tens of nanometers along the surface normal and can be used to excite fluorophores in a defined sample area in a selective way. The wave decays exponentially in vertical direction and therefore is only able to excite fluorophores in close proximity to the substrate (up to about 200 nm). As a consequence, the fluorescence background is reduced and a high-resolution fluorescence image of the immediate vicinity of the substrate can be obtained [38]. The combined AFM-TIRF technique has been successfully applied to monitor the dynamics of cell adhesion [39, 40]. Cells adhere to affinity substrates by developing so-called focal adhesion points, arrangements of proteins that act as anchors. TIRF microscopy can map the location of these adhesion points while the AFM cantilever exerts forces on certain cell positions. In this way, force transmission from the apical membrane to the basal membrane of cells has been detected from variations in the number and arrangement of focal adhesion points in TIRM images [39]. Although the integrated AFM-TIRF was well established almost two decades ago, its application in bacterial studies did not get enough interests from researchers compare to mammalian cells studies [41].

1.6.3   INTEGRATED AFM/RAMAN SPECTROSCOPY

Raman spectroscopy is a suitable method for investigating vibrational and rotational states of chemical bonds and symmetry of molecules [42]. Tip-Enhanced Raman Spectroscopy (TERS) is a near-field microscopy that makes use of an apertureless probe to enhance the Raman signal exerted by a sample (Figure 1.9). The probe is a sharp metallic tip, typically a metal-coated AFM tip or an electrochemically etched metal wire, that is irradiated along its apical axis by a low-power laser at a visible wavelength (500–650 nm). When this tip is placed sufficiently close (approx. 1 nm) to the sample, field-enhancement occurs leading to molecular excitation of the nearby sample area and thus local Raman spectra can be obtained. The combination of AFM-TERS, offers high spatial resolution of the scanning probe microscopies with surface-enhanced Raman spectroscopy suitable for structural investigations at single-molecule level for biological and biotechnological applications [33, 43, 44]. The chemical enhancement (also called charge transfer) is based on molecular interactions between the metal tip and the sample surface that alters the spectroscopic properties of the latter. In this regard, TERS presents an additional advantage with respect to the classical bulk or surface-enhanced Raman spectroscopy since Raman spectra can even be obtained from poor Raman scatterers, chemical specimens that otherwise exhibit very low or undetectable Raman signals [33].

TERS has been validated through its application to model systems such as dye molecules and carbon nanotubes, which both exhibit characteristic and well-known Raman signals [45]. As far as biomaterials are concerned, studies have been accomplished on nucleic acids and biofilms such as alginates or bacterial surfaces [46]. Wood et al. [47] demonstrated that TERS can be used to investigate hemozoin crystals at less than 20-nm spatial resolution in a digestive vacuole of a sectioned malaria parasite-infected erythrocyte cell. The characteristic Raman scattering of hemozoin, a five-coordinate high-spin ferric heme complex, could be assigned to a precise position inside a sectioned single infected cell, measured with AFM. Cialla et al. [48] showed that single (tobacco mosaic) virus could be observed with such technique, obtaining vibrational spectroscopic information with a spatial resolution less than 50 nm. On the other hand, the application of TERS to complex biological samples such as biofilms is still cumbersome work that involves proper spectral interpretation and makes it difficult to locate characteristic bands that are sample-specific [46].

1.7   CONCLUDING REMARKS

Over the past years, AFM-based methods have undoubtedly had a considerable impact in characterizing and manipulating a wide range of biological and synthetic biointerfaces ranging from tissues, cells, membranes, proteins, nucleic acids, and functional materials. In this chapter, we tried to highlight the most popular AFM-based modalities that have been implemented over the years leading to the multiparametric and multifunctional characterization of bacterial surfaces. The outcome of AFM studies is dependent on several key factors, including the nature and quality of sample, the quality of probe used and methods for modifying/characterizing probe modifications, careful control of experimental conditions, the accuracy of data collection and robust methods for data interpretation, and attention to potential artifact. Bacteria generally must be immobilized for scanning in an aqueous environment. Although there is continuous development in immobilization protocols yet most of gram-negative species are still challenging to investigate with AFM. The main challenge here is that fixation of bacteria either on hard surface or on the AFM probe may change the properties of bacterial surfaces in ways different than in the biological environment.

In force spectroscopy measurements, a modified probe may encounter the sample hundreds and even thousands of times, a situation vastly different than found in vivo where it is usually an initial interaction that provides the impetus to biological response. Simple experimental factors such as probe loading force, loading rates, scanning rates, and contact time are all important parameters that affect the interaction forces and therefore may skew outcomes [49, 50]. Experience is often necessary to understand how these factors affect the measured data.

Although there are clearly important considerations to be kept in mind for all AFM work, the field has seen, and will continue to see, an increased use of novel, high-resolution AFM techniques for microbial studies. The future applications of AFM in microbiology and life science are continuing to expand and have not yet reached their full potential. Integration of AFM with optical imaging and spectroscopic techniques [51, 52] has greatly expanded the amount and types of data one can obtain simultaneously. AFM has become a routine surface characterization technique in many laboratories and research centers, and there has been a rapidly increasing use of AFM for characterization of the biological processes including cell-cell and cell-material interactions. With the development of micromachining and nanofabrication technologies, ultra-sharp, ultrasensitive probes are being used more routinely, allowing more access to data that formerly required devoted facilities. As new developments in instrumentation and accessories continue to appear, it is likely that the versatility of the AFM as a standard tool in life sciences will continue to progress as well.

1.8   REFERENCES

[1] Binnig, G. and Rohrer, H. Scanning tunneling microscopy Scanning tunneling microscopy. *Helv. Phys. Acta*, 55:726–735, 1982. 1

[2] Goldie, I. and Wetterqvist, H. Plethysmographic and intramedullary pressure measurements before and after tibial osteotomy for osteoarthritis of the knee. *Acta Orthop. Belg.*, 40:285–293, 1974. 2

[3] Israelachvili, J. *Intermolecular and Surface Forces*, 2015. DOI: 10.1016/C2009-0-21560-1. 2, 9

[4] Baykara, M. Z. and Schwarz, U. D. *Atomic Force Microscopy: Methods and Applications*. Encyclopedia of Spectroscopy and Spectrometry, Elsevier Ltd., 2016. DOI: 10.1016/b978-0-12-409547-2.12141-9. 3, 5, 6

[5] Pillet, F., Chopinet, L., Formosa, C., and Dague, É. Atomic force microscopy and pharmacology: From microbiology to cancerology. *Biochim. Biophys. Acta—Gen. Subj.*, 1840:1028–1050, 2014. DOI: 10.1016/j.bbagen.2013.11.019. 3

[6] Raczkowska, J. et al. Structure evolution in layers of polymer blend nanoparticles. *Langmuir*, 23:7235–7240, 2007. DOI: 10.1021/la062844n. 5

[7] Heinz, W. F. and Hoh, J. H. Spatially resolved force spectroscopy of biological surfaces using the atomic force microscope. *Trends Biotechnol.*, 17:143–150, 1999. DOI: 10.1016/s0167-7799(99)01304-9. 6, 9

[8] Carvalho, F. A. and Santos, N. C. Atomic force microscopy-based force spectroscopy—Biological and biomedical applications. *IUBMB Life*, 64:465–472, 2012. DOI: 10.1002/iub.1037. 6

[9] Efremov, Y. M., Wang, W. H., Hardy, S. D., Geahlen, R. L., and Raman, A. Measuring nanoscale viscoelastic parameters of cells directly from AFM force-displacement curves. *Sci. Rep.*, 7, 2017. DOI: 10.1038/s41598-017-01784-3. 9

[10] Hecht, F. M. et al. Imaging viscoelastic properties of live cells by AFM: Power-law rheology on the nanoscale. *Soft Matter*, 11:4584–4591, 2015. DOI: 10.1039/c4sm02718c. 9, 11

[11] Moeendarbary, E. et al. The cytoplasm of living cells behaves as a poroelastic material. *Nat. Mater.*, 12:253–261, 2013. DOI: 10.1038/nmat3517. 11

[12] Stewart, M. P. et al. Hydrostatic pressure and the actomyosin cortex drive mitotic cell rounding. *Nature*, 469:226–231, 2011. DOI: 10.1038/nature09642. 11

[13] Ramanathan, S. P. et al. Cdk1-dependent mitotic enrichment of cortical myosin II promotes cell rounding against confinement. *Nat. Cell Biol.*, 17:148–159, 2015. DOI: 10.1038/ncb3098. 11

[14] Fabry, B. et al. Scaling the microrheology of living cells. *Phys. Rev. Lett.*, 87, 2001. DOI: 10.1103/physrevlett.87.148102. 11

[15] Fischer-Friedrich, E. et al. Rheology of the active cell cortex in mitosis. *Biophys. J.*, 111:589–600, 2016. DOI: 10.1016/j.bpj.2016.06.008. 11

[16] Chizhik, S. A., Huang, Z., Gorbunov, V. V., Myshkin, N. K., and Tsukruk, V. V. Micromechanical properties of elastic polymeric materials as probed by scanning force microscopy. *Langmuir*, 39:1146–1147, ACS, 1998. DOI: 10.1021/la980042p. 11, 12, 13

[17] Tsukruk, V. V., Huang, Z., Chizhik, S. A., and Gorbunov, V. V. Probing of micro mechanical properties of compliant polymeric materials. *J. Mater. Sci.*, 33:4905–4909, 1998. DOI: 10.1023/A:1004457532183. 11

[18] Oliver, W. C. and Brotzen, F. R. On the generality of the relationship among contact stiffness, contact area, and elastic modulus during indentation. *J. Mater. Res.*, 7:613–617, 1992. DOI: 10.1557/jmr.1992.0613. 11

[19] Hansen, J. C. et al. Effect of surface nanoscale topography on elastic modulus of individual osteoblastic cells as determined by atomic force microscopy. *J. Biomech.*, 40:2865–2871, 2007. DOI: 10.1098/rspa.1971.0141. 13

[20] Johnson, K. L., Kendall, K., and Roberts, A. D. Surface energy and the contact of elastic solids. *Proc. R. Soc. A Math. Phys. Eng. Sci.*, 324:301–313, 1971. DOI: 10.1098/rspa.1971.0141. 13

[21] Derjaguin, B., Muller, V., and Toporov, Y. Effect of contact deformation on the adhesion of elastic solids. *J. Colloid Interf. Sci.*, 53:314–326, 1975. DOI: 10.1016/0021-9797(75)90018-1. 13

[22] Shi, X. and Zhao, Y. P. Comparison of various adhesion contact theories and the influence of dimensionless load parameter. *J. Adhes. Sci. Technol.*, 18:55–68, 2004. DOI: 10.1163/156856104322747009. 13

[23] Avila, A. and Bhushan, B. Electrical measurement techniques in atomic force microscopy. *Crit. Rev. Solid State Mater. Sci.*, 35:38–51, 2010. DOI: 10.1080/10408430903362230. 14

[24] Kondo, Y., Osaka, M., Benten, H., Ohkita, H., and Ito, S. Electron transport nanostructures of conjugated polymer films visualized by conductive atomic force microscopy. *ACS Macro Lett.*, 4:879–885, 2015. DOI: 10.1021/acsmacrolett.5b00352. 14

[25] Tararam, R., Garcia, P. S., Deda, D. K., Varela, J. A., and de Lima Leite, F. *Atomic Force Microscopy: A Powerful Tool for Electrical Characterization*, Nanocharacterization Techniques, Elsevier Inc., 2017. DOI: 10.1016/b978-0-323-49778-7.00002-3. 14

[26] Cleveland, J. P., Manne, S., Bocek, D., and Hansma, P. K. A nondestructive method for determining the spring constant of cantilevers for scanning force microscopy. *Rev. Sci. Instrum.*, 64:403–405, 1993. DOI: 10.1063/1.1144209. 15

[27] Eghiaian, F., Rico, F., Colom, A., Casuso, I., and Scheuring, S. High-speed atomic force microscopy: Imaging and force spectroscopy. *FEBS Lett.*, 588:3631–3638, 2014. DOI: 10.1016/j.febslet.2014.06.028. 15, 16

[28] Smith, C. Microscopy: Two microscopes are better than one. *Nature*, 492:293–297, 2012. DOI: 10.1038/492293a. 16

[29] Zhou, L., Cai, M., Tong, T., and Wang, H. Progress in the correlative atomic force microscopy and optical microscopy. *Sensors*, 17, 2017. DOI: 10.3390/s17040938. 16, 17

[30] König, M., Koberling, F., Gmbh, P., Walters, D., and Viani, J. Combining atomic force microscopy with confocal microscopy. *Microscopy*, pages 1–6, 2009.

[31] Geisse, N. A. AFM and combined optical techniques. *Mater. Today*, 12:40–45, 2009. DOI: 10.1016/s1369-7021(09)70201-9.

[32] Harke, B., Chacko, J. V., Haschke, H., Canale, C., and Diaspro, A. A novel nanoscopic tool by combining AFM with STED microscopy. *Opt. Nanosc.*, 1:1–6, 2012. DOI: 10.1186/2192-2853-1-3. 17

[33] Kainz, B., Oprzeska-Zingrebe, E. A., and Herrera, J. L. Biomaterial and cellular properties as examined through atomic force microscopy, fluorescence optical microscopies and spectroscopic techniques. *Biotechnol. J.*, 9:51–60, 2014. DOI: 10.1002/biot.201300087. 17, 19

[34] Ramires, P. A., Giuffrida, A., and Milella, E. Three-dimensional reconstruction of confocal laser microscopy images to study the behaviour of osteoblastic cells grown on biomaterials. *Biomaterials*, 23:397–406, 2002. DOI: 10.1016/s0142-9612(01)00118-1. 18

[35] Timmel, T., Schuelke, M., and Spuler, S. Identifying dynamic membrane structures with atomic-force microscopy and confocal imaging. *Microsc. Microanal.*, 20:514–520, 2014. DOI: 10.1017/s1431927613014098. 18

[36] Bhat, S. V. et al. Correlative atomic force microscopy quantitative imaging-laser scanning confocal microscopy quantifies the impact of stressors on live cells in real-time. *Sci. Rep.*, 8, 2018. DOI: 10.1038/s41598-018-26433-1. 18

[37] Hinterdorfer, P., Baumgartner, W., Gruber, H. J., Schilcher, K., and Schindler, H. Detection and localization of individual antibody-antigen recognition events by atomic force microscopy. *Proc. Natl. Acad. Sci.*, 93:3477–3481, 1996. DOI: 10.1073/pnas.93.8.3477. 18

[38] Axelrod, D., Burghardt, T. P., and Thompson, N. L. Total internal reflection fluorescence. *Annu. Rev. Biophys. Bioeng.*, 13:247–268, 1984. DOI: 10.1146/annurev.bb.13.060184.001335. 19

[39] Mathur, A. B., Truskey, G. A., and Reichert, W. M. Atomic force and total internal reflection fluorescence microscopy for the study of force transmission in endothelial cells. *Biophys. J.*, 78:1725–1735, 2000. DOI: 10.1016/s0006-3495(00)76724-5. 19

[40] Trache, A. and Lim, S. M. Live cell response to mechanical stimulation studied by integrated optical and atomic force microscopy. *J. Vis. Exp.*, 2010. DOI: 10.3791/2072. 19

[41] Brown, A. E. X., Hategan, A., Safer, D., Goldman, Y. E., and Discher, D. E. Cross-correlated TIRF/AFM reveals asymmetric distribution of force-generating heads along self-assembled, "synthetic1" myosin filaments. *Biophys. J.*, 96:1952–1960, 2009. DOI: 10.1016/j.bpj.2008.11.032. 19

[42] Stiles, P. L., Dieringer, J. A., Shah, N. C., and Van Duyne, R. P. Surface-enhanced raman spectroscopy. *Annu. Rev. Anal. Chem.*, 1:601–626, 2008. DOI: 10.1146/annurev.anchem.1.031207.112814. 19

[43] Deckert-Gaudig, T. and Deckert, V. Nanoscale structural analysis using tip-enhanced Raman spectroscopy. *Curr. Opin. Chem. Biol.*, 15:719–724, 2011. DOI: 10.1016/j.cbpa.2011.06.020. 19

[44] Elfick, A. P. D., Downes, A. R., and Mouras, R. Development of tip-enhanced optical spectroscopy for biological applications: A review. *Analyt. Bioanalyt. Chem.*, 396:45–52, 2010. DOI: 10.1007/s00216-009-3223-9. 19

[45] Domke, K. F. and Pettinger, B. Studying surface chemistry beyond the diffraction limit: 10 years of TERS. *ChemPhysChem*, 11:1365–1373, 2010. DOI: 10.1002/cphc.200900975. 19

[46] Cialla, D. et al. Surface-enhanced Raman spectroscopy (SERS): Progress and trends. *Analyt. Bioanalyt. Chem.*, 403:27–54, 2012. DOI: 10.1007/s00216-011-5631-x. 19, 20

[47] Wood, B. R. et al. Tip-enhanced raman scattering (TERS) from hemozoin crystals within a sectioned erythrocyte. *Nano Lett.*, 11:1868–1873, 2011. DOI: 10.1021/nl103004n. 19

[48] Cialla, D. et al. Raman to the limit: Tip-enhanced Raman spectroscopic investigations of a single tobacco mosaic virus. *J. Raman Spectrosc.*, 40:240–243, 2009. DOI: 10.1002/jrs.2123. 20

[49] Xu, L. C., Vadillo-Rodriguez, V., and Logan, B. E. Residence time, loading force, pH, and ionic strength affect adhesion forces between colloids and biopolymer-coated surfaces. *Langmuir*, 21:7491–7500, 2005. DOI: 10.1021/la0509091. 20

[50] Dufrêne, Y. F. Atomic force microscopy and chemical force microscopy of microbial cells. *Nat. Protoc.*, 3:1132–1138, 2008. DOI: 10.1038/nprot.2008.101. 20

[51] Kassies, R. et al. Combined AFM and confocal fluorescence microscope for applications in bio-nanotechnology. *J. Microsc.*, 217:109–116, 2005. DOI: 10.1111/j.0022-2720.2005.01428.x. 20

[52] Schmidt, U., Hild, S., Ibach, W., and Hollricher, O. Characterization of thin polymer films on the nanometer scale with confocal Raman AFM. *Macromol. Symp.*, 230:133–143, 2005. DOI: 10.1002/masy.200551152. 20

CHAPTER 2

Protocols for Microbial Specimen Preparation for AFM Analysis

OVERVIEW

Investigating dynamic cellular processes in microbial cells in their native, aqueous environment is important for understanding the architectural dynamics, surface molecular interactions, etc., of the microbes. The medical and clinical relevance of such studies have spurred intense research, with a considerable focus in recent times on using the high-resolution capability of AFM to probe the microbial world. However, AFM imaging of bacterial cells in liquid in order to visualize live cells performing natural cellular processes has proved difficult. Although the potential of the AFM for imaging animal cells was recognized quite early, attempts to perform such measurements on microbes have been more limited, due to a large extent of difficulties associated with sample preparation. Unlike animal cells, microbes have a well-defined shape and have no tendency to spread on surfaces. As a result, the contact area between a cell and a support is very small, often leading to cell detachment by the scanning tip. In addition, the bacterial surface is the binding interface, and given the inherent differences in bacterial surface properties, the methodology of sample preparation must be determined for each specimen. Therefore, several strategies have been developed to promote firm attachment of the cells. This process includes selecting the form of immobilization (i.e., physical vs. chemical), and in live cell imaging, identifying physiologically compatible immobilization and imaging buffers. Both aspects must be carefully chosen based on the experimental goals and the data to be collected. Specifically, the strategy of immobilization must not alter the cell properties being studied, and the cells must be oriented in such a fashion that interactions between the probe and the region of interest are unimpeded.

The development of immobilization methods, and subsequent improvement of those methods, has continued over the last 20 years for which there is a correlating increase in the application of AFM in microbial surface studies. This chapter reviews the most common methods of sample preparation that are used for imaging and manipulating microbial cells with an AFM.

2.1   PHYSICAL IMMOBILIZATION

2.1.1   PORE TRAPPING METHOD

There are many protocols for chemical immobilization of microbial cells but very few for physical immobilization. The shape and the outer structure of the cell are the main factors that impose the choice of the physical immobilization technique. An established method of immobilizing coccoid bacteria is to trap cells in polycarbonate track etched filter pores [1–3]. This method was first adapted by Kasas and Ikai (in 1995), to image *Saccharomyces cerviceae* cells [1]. As described in Figure 2.1, the technique is very simple and consists in filtering a small bacteria suspension using polymeric membranes with pore sizes comparable to the dimensions of the cells. The cells become mechanically trapped in the pores, allowing repeated imaging without cell detachment or damage. This trapping method has been successfully used to immobilize a number of bacterial strains and species including fibrillated (strain HB) and non-fibrillated (strain HBC12) *Streptococcus salivarius* [4], *Streptococcus mutans* [5], *Staphylococcus aureus* [2, 6, 7], *Staphylococcus epidermis* [8, 9], and *Lactococcus lactis* [10]. Pore trapping is, in principle, less likely to affect the native state of organisms than chemical fixation [8, 11]. The rigidity provided by the peptidoglycan layer in the cell assists in maintaining the spherical conformation of the bacteria as well as holding the cells in the pores (Figure 2.1b). However, the diameter of the organisms to be trapped and the size of the filter pores in which they were to be immobilized have generally not been known and availability of filters has been limited to those which can be purchased commercially. This seriously restricts the versatility of the technique with respect to the range of organisms to which it might potentially be applied and may mean that organisms are trapped at low surface densities requiring the AFM operator to spend many hours searching for trapped organisms; making comparative measurements between cells from the same culture at the same phase very difficult. To improve the pore trapping technique, Turner and coworkers [3] etched commercially available polymeric filters to a better size for trapping, making the method more effective and potentially applicable to a wider range of strains and species.

Although the pore trapping method is generally not applicable for immobilizing rod-shaped bacteria or cell wall deficient forms of bacteria, at least one report applied the technique for imaging rod-shaped mycobacteria [12].

Similar methodology, but less effective, was developed by Meyer et al. [13]. It consists in physical confinement of cells in microwells of modified glass surfaces. This technique is less effective and based on random trap of cells in 0.5 μm deep and 1.5 μm wide microwells made by colloid lithography [13]. A colloidal monolayer mask is deposited on a clean microscope glass by electrostatic self-assembly. A thin gold film (about 600–700 nm) is then deposited by electron-beam evaporation. Finally, tape is gently placed on the thin gold film, and the dispersed colloids on the glass substrate are removed by pulling off the tape, exposing the microwells. Bacteria can be immobilized in the microwells by placing a drop of bacterial solution on the slide and incubating for 20 min before washing gently the slide. The success of cell capture is not very reproducible and large area need to be scanned to locate cells by an AFM. The frequency of

(a)

(b)

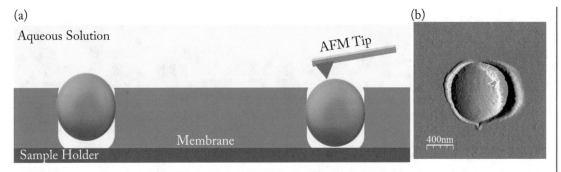

Figure 2.1: (a) Schematic representation of the pore trapping immobilization method. Cells are physically trapped into porous polymeric membrane. (b) A deflection AFM image of a trapped *Staphylococcus aureus* cell imaged under aqueous condition.

capturing cells in the wells may be increased by modifying the chemistry inside wells to combine physical entrapment with electrostatic interactions [14].

2.1.2 MICROFLUIDIC BACTERIAL TRAPS

Recently, a microfluidic device was developed to physically trap bacteria with any shape and size [15]. This device can be used to easily immobilize bacteria in well-defined positions and subsequently release the cells for quick sample exchange. The microfluidic chip assembly compounds are fabricated from chemically inert materials, as shown in Figure 2.2a. The bacteria are attracted toward the traps and physically immobilized by creating a pressure difference across the nanofluidic traps (Figure 2.2b). The release of the immobilized bacteria or the repel of any undesirable particles from the traps can be performed on demand by simply reversing the pressure difference. In addition, the thickness of the microfluidic chip assembly allows simultaneous fluorescence analysis to assess the bacterial viability and correlate measurements during AFM manipulations. Regarding the reusability of the microfluidic chip, the whole bacterial trapping device including all microfluidic parts can be decontaminated and cleaned after each experiment using alcohol or other cleaning agents. With this devise rapid and robust immobilization of motile (Figure 2.2c), as well as non-motile bacteria, are possible, accelerating versatile applications across many bacterial species and liquid media.

2.2 PHYSICOCHEMICAL AND COVALENT IMMOBILIZATION

2.2.1 ELECTROSTATIC IMMOBILIZATION

Substrates covered with a cationic polymer or hydrogel with a regular structure, such as gelatin, poly-L-lysine, polyethyleneimine, agar, and agarose are frequently used for immobilization of

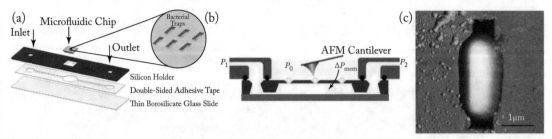

Figure 2.2: (a) Assembly of the microfluidic chip containing the bacterial traps. The microfluidic chip is mounted to the silicon holder in a central, square opening. The silicon holder is attached to a borosilicate glass slide through a prepatterned double-sided adhesive tape to form the microfluidic channels. (b) Measurement configuration for bacterial trapping using AFM microscopy setup. A pressure difference ΔP_{mem} is applied across the membrane containing the bacterial traps. The top side is exposed to atmospheric pressure P_0, whereas the pressure in the channel is controlled through the input P_1 and output P_2. (c) AFM image of a trapped E. coli bacterium in lysogeny broth growth medium. The image was obtained in AFM tapping mode. (Reprinted from ref. [15] with permision from Springer Nature. Copyright (2017).)

microorganisms during AFM scanning [16–18]. Contrary to immobilization on membrane filters and lithographic grids, this method is suitable for microbial cells of any shape. As shown in Figure 2.3a, immobilization occurs due to the electrostatic interaction between negatively charged microbial cells and positively charged groups on the polymer surface (partly involving hydrophobic forces) [16, 19, 20]. An important requirement for the sample preparation is that the polymer or gel layer should be very thin in order to prevent the cell penetration into the gel due to the force applied by the AFM cantilever [19, 21]. The surface characteristics of the studied microorganisms, the duration of cell incubation on the polymer surface, and the presence of chemical substances in the medium affect the efficiency of immobilization [22, 23]. Thus, it was registered that buffer solutions with high ionic strength added as a liquid medium during AFM scanning of S. sciuri decreased the efficacy of electrostatic interactions between bacterial cells and a polymer-modified substrate, while the most efficient retention of bacteria was observed in deionized water [23]. In addition, the surface charge of microorganisms can be increased depending on the cultivation conditions up to a shift in the positive region which prevents the electrostatic cell attraction to the substrate covered by a cationic polymer [24]. This method of cell immobilization is widely used in studies of nanomechanical and adhesive properties of living bacteria [25, 26].

2.2.2   IMMOBILIZATION USING COVALENT BINDING

Immobilizing cells covalently to a surface by chemical fixation or using cross-linking agents may be required either to avoid microbial displacement by the scanning AFM tip or to obtain high-

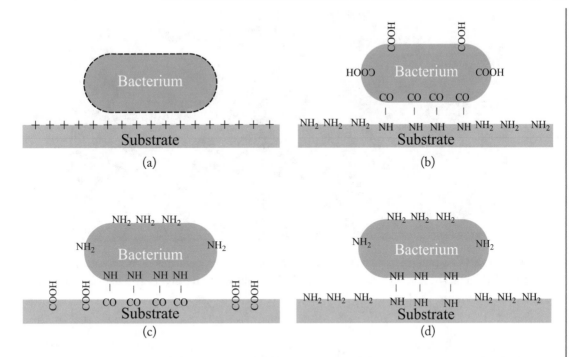

Figure 2.3: Schematic representation of the immobilization methods employed. (a) Attractive electrostatic interactions, (b) covalent binding to amine-functionalized surfaces, (c) covalent binding to carboxyl-functionalized surfaces, and (d) covalent binding to amine functionalized surfaces.

resolution images. This method enables bacteria to be bound either to surfaces or to AFM tips enabling a range of measurements on mechanical and adhesive properties of organisms to be obtained that would otherwise be unavailable. As illustrated in Figure 2.3b,c,d, there are several methods of immobilization of bacterial cells based on the formation of covalent bonds with the substrate surface via carboxyl groups or amino groups. The covalent functionalization of bacteria is usually preferable for gram-negative bacteria that is hard to immobilized by other means. In addition, this type of bacteria fixation provides a strong attachment of cells and does not affect significantly bacteria growth and division during AFM measurements under physiological conditions [27]. For example, *Burkholderia cepacia* and *Pseudomonas stutzeri* cells were immobilized on silanized glasses treated with 1-ethyl-3-(3-dimethylaminopropyl) carbodiimide (EDC) and N-hydroxysuccinimide (NHS) solutions. AFM imaging immediately after bacteria treatment with linking reagents showed that only low level ($\leq$ 15%) of cell death and cell-wall disruption [28]. However, the commonly active groups used in covalent binding (polyethyleneimine or poly-L-lysine) and the reagents used for cross-linking, such as glutaraldehyde, have been

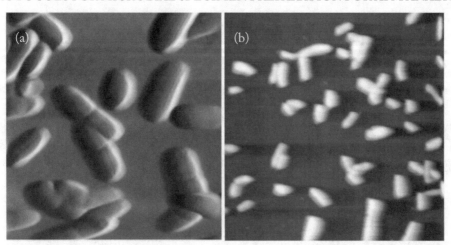

Figure 2.4: (a) AFM images of *E. coli*. The cells are mounted on gelatin-coated mica and imaged in 0.005 M PBS22. (b) AFM image of *E. coli* cells immobilized on poly-L-lysine in 0.01 M PBS. (a) Adapted from ref. [22].

shown to affect cell viability through permeabilization of cell membranes or by cross-linking the proteins on the cell surface. Meyer et al. [27] compared two different ways of covalently immobilization bacteria for AFM imaging in a liquid. In the first protocol, a cellular suspension treated with EDC and NHS was placed on glass surface modified with polyacrylic acid, EDC, or poly-L-lysine (PLL). The second method consists in placing a cellular suspension on glass surface modified with glutaraldehyde. The most important finding of the study is that the possibility of the chemical modification of cells using the second way is minimal: most cells remain viable and strongly attached to the substrate during the scanning in a liquid medium which allows obtaining of high-quality AFM images.

2.2.3 IMMOBILIZATION USING ADHESIVE PROTEINS, LECTINS, AND ANTIBODIES

Because of the challenges in immobilizing bacteria species for AFM measurements, scientists are always trying to come up with new and specific protocols that preserve bacterial activity and simultaneously allow AFM measurements under physiological conditions. In this context, several studies were reported in using proteins, lectins, and antibodies to covalently attach the bacteria to hard surfaces. For example, an extract of polyphenol proteins, secreted by the marine Mytilus edulis, was used for the attachment of bacterial cells with different shapes and sizes for AFM studies in aqueous conditions (Figure 2.5) [29]. An extract of these adhesive proteins are now available as a commercial product Cell-Tak™ and this method provided a secure attachment of bacterial cells on the substrate. The viability of the immobilized bacteria

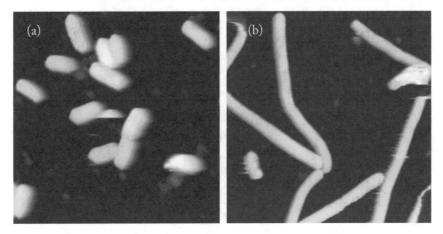

Figure 2.5: (a) *E. coli*, (b) *B. subtilis* bacteria imaged by contact mode in 20 mM HEPES after immobilization by an attachment to polyphenolic adhesive proteins.

as well as their mechanical properties are not affected by the presence of the protein. Studies of these functionally unique proteins have revealed that the adhesive effect of this protein is due to the presence of the unusual amino acid 3,4-dihydroxy-L-phenyla-lanine (DOPA) [29, 30]. A biomimetic adhesive polymer called polydopamine was inspired from the chemical composition of these proteins [30]. AFM studies using the polydopamine to attach bacteria to hard surfaces, showed no effect on cell viability and cell surface properties [30, 31]. Also, this sticky polymer was demonstrated to be suitable for immobilizing bacterial AFM cantilevers for AFM force measurements studies [31, 32]. This method was proven to provide a gentle and consistent immobilization, which can be used for bacteria of all shapes and sizes, and does not require exposure to chemicals, distilled water, or drying—all of which may affect cell viability, cell surface properties, or trigger an undesirable biological response.

Lectin functionalizing hard surfaces was also proven to be an alternative promising immobilization technique for living microbial cells. This method was first developed by E. Dague et al. for the attachment of *S. cerevisiae* cells and *Aspergillus fumigatus* spores to microstructured lithographic grids made from polydimethylsiloxane (PDMS) [33]. Functionalization of the PDMS surface was performed using a combination of UV and ozone or by the treatment with a plant lectin (concanavalin A). Yeast cells and fungal spores were immobilized on a PDMS grid by convective-capillary precipitation from the suspension at a constant rate of drop movement along the sample and scanned in a liquid medium. The data obtained by the authors indicate the advantage of the biological functionalization of the substrate (with lectins) over physicochemical activation, which was proven by a series of high-quality AFM images and reproducible force curves obtained for yeast cells. However, the use of lectins for microorganism immobiliza-

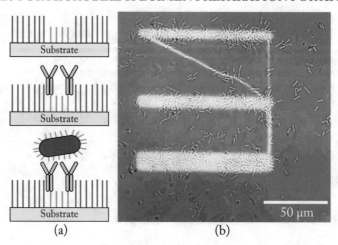

Figure 2.6: (a) The substrate was first modified with chemicals that inhibit the nonspecific adsorption of proteins (blue bars) and then etched using a programmable focused Ga+ ion beam. The freshly etched surface was then modified with a cross-linker (orange bars) to link the antibody (cyan Y shapes), raised against the bacterial surface antigens, to the substrate. When the patterned substrate was incubated with the bacterial suspension, the bacterial cells adhered only to the antibody-modified area and thus formed a monolayer of bacterial cell patterns. (b) Micropatterns of live S. Typhimurium cells immobilized on substrates etched by a focused ion beam (line pattern on silicon). Adapted with permission from ref. [27]. Copyright (2008) American Chemical Society.

tion is limited to cells that have certain sugars on the surface (that are specific lectin-binding sites) [33, 34].

An original method of specific immobilization of living bacteria using poly- and monoclonal antibodies was suggested in a series of studies by Suo et al. [35, 36]. Antibodies to Salmonella enterica fimbria CFA/I antigens were noncovalently adsorbed on the surface of a silicone substrate, and then a bacterial suspension was applied on the modified substrate, where their specific binding to antibodies occurred [36]. The immunoadsorption efficiencies of salmonellae when using antibodies to CFA/I antigens of fimbriae, flagella, and lipopolysaccharides (LPS) and the capsule F1 antigen were compared [35]. The cell adsorption to monoclonal antibodies against CFA/I fimbriae, as well as against flagellin of flagella (only in those cases when the mobility of flagella was genetically paralyzed) was the most efficient. Among antibodies specific to LPS, a positive result was obtained only when using O-polysaccharide antibodies. The attachment of bacteria to a surface modified with antibodies to the F1 antigen was the least efficient, presumably, due to desorption of the F1 protein from the cell surface. Thus, the selected antigen-antibody pairs for immunospecific adsorption of bacteria allow one to attach cells on a flat substrate without any effect on their viability and surface properties. A large variety of surface antigens (and, accordingly, antibodies) opens wide possibilities for selective immobilization of microorganisms by taxonomic or functional traits. It should be noted that the described methods were developed by the authors in order to create biosensors based on microchips with attached bacterial cells; however, immobilization of microorganisms via an antigen-antibody reaction can be used during sample preparation for AFM. The difficulties in obtaining and purification of monoclonal antibodies (as well as the low efficiency of microbial cell immobilization in some cases) are the disadvantages of the method.

To summarize, the above methods of microorganism attachment to flat substrates for AFM in a liquid medium have certain advantages and disadvantages. Therefore, the selection of an optimal immobilization method depends on the particular task and morphological-physiological peculiarities of the microbial cells.

2.3 REFERENCES

[1] Kasas, S. and Ikai, A. A method for anchoring round shaped cells for atomic force microscope imaging. *Biophys. J.*, 68, 1995. DOI: 10.1016/s0006-3495(95)80344-9. 28

[2] Touhami, A., Jericho, M. H., and Beveridge, T. J. Atomic force microscopy of cell growth and division in staphylococcus aureus. *J. Bacteriol.*, 186, 2004. DOI: 10.1128/jb.186.11.3286-3295.2004. 28

[3] Turner, R. D., Thomson, N. H., Kirkham, J., and Devine, D. Improvement of the pore trapping method to immobilize vital coccoid bacteria for high-resolution AFM: A study of staphylococcus aureus. *J. Microsc.*, 238:102–110, 2010. DOI: 10.1111/j.1365-2818.2009.03333.x. 28

[4] van der Mei, H. C. et al. Direct probing by atomic force microscopy of the cell surface softness of a fibrillated and nonfibrillated oral streptococcal strain. *Biophys. J.*, 78(5):2668–2674, 2000. DOI: 10.1016/s0006-3495(00)76810-x. 28

[5] Busscher, H. J. et al. Intermolecular forces and enthalpies in the adhesion of streptococcus mutans and an antigen I/II-deficient mutant to Laminin films. *J. Bacteriol.*, 189:2988–2995, 2007. DOI: 10.1128/jb.01731-06. 28

[6] Francius, G., Domenech, O., Mingeot-Leclercq, M. P., and Dufrêne, Y. F. Direct observation of staphylococcus aureus cell wall digestion by lysostaphin. *J. Bacteriol.*, 190:7904–7909, 2008. DOI: 10.1128/jb.01116-08. 28

[7] Turner, R. D., Kirkham, J., Devine, D., and Thomson, N. H. Second harmonic atomic force microscopy of living Staphylococcus aureus bacteria. *Appl. Phys. Lett.*, 94, 043901, 2009. DOI: 10.1063/1.3073825. 28

[8] Vadillo-Rodríguez, V. et al. Comparison of atomic force microscopy interaction forces between bacteria and silicon nitride substrata for three commonly used immobilization methods. *Appl. Environ. Microbiol.*, 70:5441–5446, 2004. DOI: 10.1128/aem.70.9.5441-5446.2004. 28

[9] Méndez-Vilas, A., Gallardo-Moreno, A. M., Calzado-Montero, R., and González-Martín, M. L. AFM probing in aqueous environment of Staphylococcus epidermidis cells naturally immobilised on glass: Physico-chemistry behind the successful immobilisation. *Colloids Surf. B Biointerf.*, 63:101–109, 2008. DOI: 10.1016/j.colsurfb.2007.11.011. 28

[10] Gilbert, Y. et al. Single-molecule force spectroscopy and imaging of the Vancomycin/D-Ala-D-Ala interaction. *Nano. Lett.*, 19:25, 2019. DOI: 10.1021/nl0700853. 28

[11] Ahimou, F., Denis, F. A., Touhami, A., and Dufrêne, Y. F. Probing microbial cell surface charges by atomic force microscopy. *Langmuir*, 18, 2002. DOI: 10.1021/la026273k. 28

[12] Dupres, V. et al. Nanoscale mapping and functional analysis of individual adhesins on living bacteria. *Nat. Meth.*, 2:515, 2005. DOI: 10.1038/nmeth769. 28

[13] Louise Meyer, R. et al. Immobilisation of living bacteria for AFM imaging under physiological conditions. *Ultramicroscopy*, 110:1349–1357, 2010. DOI: 10.1016/j.ultramic.2010.06.010. 28

[14] Rowan, B., Wheeler, M. A., and Crooks, R. M. Patterning bacteria within hyperbranched polymer film templates. *Langmuir*, 18(25): 9914–9917, 2002. DOI: 10.1021/la020664h. 29

[15] Peric, O., Hannebelle, M., Adams, J. D., Fantner, G. E., and Berlin, S.-V. Microfluidic bacterial traps for simultaneous fluorescence and atomic force microscopy. DOI: 10.1007/s12274-017-1604-5. 29, 30

[16] Bolshakova, A. V. et al. Comparative studies of bacteria with an atomic force microscopy operating in different modes. *Ultramicroscopy*, 86:121–128, 2001. DOI: 10.1016/s0304-3991(00)00075-9. 30

[17] Doktycz, M. J. et al. AFM imaging of bacteria in liquid media immobilized on gelatin coated mica surfaces. *Ultramicroscopy*, 97:209–216, 2003. DOI: 10.1016/s0304-3991(03)00045-7.

[18] Velegol, S. B. and Logan, B. E. Contributions of bacterial surface polymers, electrostatics, and cell elasticity to the shape of AFM force curves. *Langmuir*, 18(13): 5256–5262, 2002. DOI: 10.1021/la011818g. 30

[19] Kuyukina, M. S., Korshunova, I. O., Rubtsova, E. V., and Ivshina, I. B. Methods of microorganism immobilization for dynamic atomic-force studies (review). *Appl. Biochem. Microbiol.*, 50:1–9. DOI: 10.1134/s0003683814010086. 30

[20] Vadillo-Rodríguez, V. et al. Comparison of atomic force microscopy interaction forces between bacteria and silicon nitride substrata for three commonly used immobilization methods. *Appl. Environ. Microbiol.*, 70:5441–5446, 2004. DOI: 10.1128/aem.70.9.5441-5446.2004. 30

[21] Kailas, L. et al. Immobilizing live bacteria for AFM imaging of cellular processes. *Ultramicroscopy*, 109:775–780, 2009. DOI: 10.1016/j.ultramic.2009.01.012. 30

[22] Allison, D. P., Sullivan, C. J., Mortensen, N. P., Retterer, S. T., and Doktycz, M. Bacterial immobilization for imaging by atomic force microscopy. *J. Vis. Exp.*, (54), e2880, 2011. DOI: 10.3791/2880. 30, 32

[23] Louise Meyer, R. et al. Immobilisation of living bacteria for AFM imaging under physiological conditions. *Ultramicroscopy*, 110:1349–1357, 2010. DOI: 10.1016/j.ultramic.2010.06.010. 30

[24] De Carvalho, C. C. C. R., Wick, L. Y., and Heipieper, H. J. Cell wall adaptations of planktonic and biofilm Rhodococcus erythropolis cells to growth on C5 to C16 n-alkane hydrocarbons. *Appl. Microbiol. Biotechnol.*, 82:311–320, 2009. DOI: 10.1007/s00253-008-1809-3. 30

[25] Webb, H. K., Truong, V. K., Hasan, J., Crawford, R. J., and Ivanova, E. P. Physico-mechanical characterisation of cells using atomic force microscopy— current research and methodologies. *J. Microbiol. Meth.*, 86:131–139, 2011. DOI: 10.1016/j.mimet.2011.05.021. 30

[26] Park, B. and Abu-Lail, N. I. Variations in the nanomechanical properties of virulent and avirulent Listeria monocytogenes. *Soft Matter*, 6:3898–3909, 2010. DOI: 10.1039/b927260g. 30

[27] Meyer, R. L. et al. Immobilisation of living bacteria for AFM imaging under physiological conditions. *Ultramicroscopy*, 110:1349–1357, 2010. DOI: 10.1016/j.ultramic.2010.06.010. 31, 32, 34

[28] Camesano, T. A., Natan, M. J., and Logan, B. E. Observation of changes in bacterial cell morphology using tapping mode atomic force microscopy. *Langmuir*, 16:4563–4572, 2000. DOI: 10.1021/la990805o. 31

[29] Lee, H., Scherer, N. F., and Messersmith, P. B. Single-molecule mechanics of mussel adhesion. *Proc. Natl. Acad. Sci.*, 103:12999–13003, 2006. DOI: 10.1073/pnas.0605552103. 32, 33

[30] Lee, H., Lee, B. P., and Messersmith, P. B. A reversible wet/dry adhesive inspired by mussels and geckos. *Nature*, 448:338–341, 2007. DOI: 10.1038/nature05968. 33

[31] Kang, S. and Elimelech, M. Bioinspired single bacterial cell force spectroscopy. *Langmuir*, 25:9656–9659, 2009. DOI: 10.1021/la902247w. 33

[32] Beaussart, A. et al. Single-cell force spectroscopy of probiotic bacteria. *Biophys. J.*, 104:1886–1892, 2013. DOI: 10.1016/j.bpj.2013.03.046. 33

[33] Dague, E. et al. Assembly of live micro-organisms on microstructured PDMS stamps by convective/capillary deposition for AFM bio-experiments. *Nanotechnology*, 22, 2011. DOI: 10.1088/0957-4484/22/39/395102. 33, 34

[34] Chopinet, L., Formosa, C., Rols, M. P., Duval, R. E., and Dague, E. Imaging living cells surface and quantifying its properties at high resolution using AFM in QI^{TM} mode. *Micron*, 48:26–33, 2013. DOI: 10.1016/j.micron.2013.02.003. 34

[35] Suo, Z., Yang, X., Deliorman, M., Cao, L., and Avci, R. Capture efficiency of escherichia coli in fimbriae-mediated immunoimmobilization, *Langmuir*, 28(2):1351–1359, 2011. DOI: 10.1021/la203348j. 34

[36] Suo, Z., Avci, R., Yang, X., and Pascual, D. W. Efficient immobilization and patterning of live bacterial cells, *Langmuir*, 24(8):4161–4167, 2008. DOI: 10.1021/la7038653. 34

CHAPTER 3

Cell Surface Structures at the Nanoscale

OVERVIEW

Microbial cell surfaces are highly complex and heterogeneous systems that fulfill several important functions, such as protecting cells against unfavorable environment, supporting the internal turgor pressure of the cell, imparting shape to the organism, sensing environmental stresses/stimuli, and controlling interfacial interactions and thereby biointerfacial processes (molecular recognition, cell adhesion, and aggregation) [1–4]. Although the structures and biochemical compositions of microbial cell-surface constituents are generally well-characterized, little is known about the spatial organization, assembly, conformational properties, and interactions of the individual components. This is largely due to the fact that traditional methods in microbiology focus on large population of cells or molecules, rather than on single cells and single molecules. In fact, isogenic microbial populations contain subgroups of cells which exhibit differences in physiological parameters such as growth rate, resistance to stress, and drug treatment [5]. For example, some bacteria display polar organelles, such as flagella and pili, or lack capsular material at the new poles following cell division. New appreciation for the importance of cellular heterogeneity, coupled with recent advances in technology, has driven the development of new tools and techniques for the study of individual microbial cells. Examples of single-cell technologies include fluorescence assays, flow cytometry techniques, microspectroscopic methods, mechanical, optical, and electrokinetic micromanipulations, microcapillary electrophoresis, biological microelectromechanical systems, and AFM.

During the past two decades, AFM-based techniques have been increasingly used for the multiparametric analysis of microbial cell surfaces, providing novel insight into their structure-function relationships. Compared to electron microscopy techniques, the main advantages of AFM for microbiologists are the possibility to image cellular structures at molecular resolution and under physiological conditions, the ability to monitor in situ the structural dynamics of cell surfaces in response to stress and to drugs, and the capability to measure the localization, adhesion, and mechanics of single cell wall constituents. As a result, scientists have been able to characterize microorganisms and their activities at unprecedented levels of detail. New insights into the properties of chemical signaling pathways and mechanisms behind the coordination of multicellular behaviors have been possible. The use of AFM in this field will continue to grow particularly due to its ability to study dynamic processes at the nanoscale and on live cells.

The following chapter explores the recent progress that has been made using high-resolution AFM imaging and AFM related methods to address microbial cell organization and dynamics. Although the primary focus of this chapter is on single cells, selected examples that deal with purified bacteria cell wall peptidoglycan are also presented.

3.1 ELUCIDATING OF MICROBIAL CELL SUBSTRUCTURE

As discussed in detail in Chapter 2, the most important steps for successful AFM investigations of single bacterial cells, are the selection of the right sample preparation and cell immobilization methodologies for the studied species. The growing number of papers published about cell immobilization reveals that this step persists as an important limitation to the application of AFM in microbiology. However, AFM is currently the only technique that can image the surface of a living cell at high resolution and in real time and over the past two decades, rapid progress has been made in applying AFM to investigate the cellular nanostructures at the single cell level with unprecedented resolution [6]. The ability of AFM to reveal native surface structures of pathogens at nanometer resolution and under physiological conditions offers exciting prospects in basic and applied research. In cellular microbiology, the technique has enabled researchers to unravel the surface structure of a variety of microbial species. *Staphylococcus aureus* [7, 8], *Bacillus spores* [9, 10], *Escherichia coli* [11], *Pseudomonas aeruginosa* [12, 13], *Mycobacterium bovis* [14], and *lactic acid bacteria* [15] are just a few examples of microbes that have been explored by AFM in recent years. Particularly, the technique has allowed the supramolecular organization of many cell wall constituents to be directly observed on live cells, including polysaccharides, peptidoglycan, and proteins. A typical example of such in vivo experiments is the nanoscale three-dimensional assembly of clustered proteinaceous microfibrils, termed rodlets observed on the surface of the human opportunistic pathogen *Aspergillus fumigatus*, which causes several harmful respiratory diseases. As shown in Figure 3.1, a single *Aspergillus fungal* spore can be trapped in a polycarbonate membrane (Figure 3.1a) and rodlets can be visualized on higher magnification (Figure 3.1b,c). Such rodlets are composed of hydrophobic proteins, which are of paramount importance to favor spore dispersion by air currents and mediate adherence to the human tissue causing important respiratory disease such as bronchopulmonary and invasive aspergillosis. Unprecedent good resolution, with a vertical resolution of ~ 0.5 nm and a lateral resolution of ~ 5 nm (Figure 3.1d), was obtained on this rodlets in buffer condition and directly on live cells [16, 17].

Dramatic changes of cell surface structure were observed upon germination of *Aspergillus fumigatus* conidia, the rodlet layer changing into a layer of amorphous material reflecting the underlying polysaccharides (Figure 3.2b) [18]. Interestingly, a similar amorphous morphology, devoid of nanostructures, was observed for a mutant affected in rodlet formation (Figure 3.2c), illustrating the usefulness of the technique in assessing the phenotypic characteristics of mutant strains with altered cell-wall constituents. Recent AFM studies have shown that rodlets are a

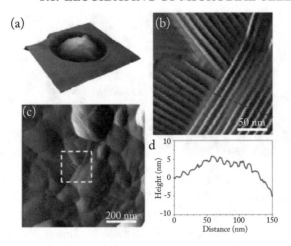

Figure 3.1: AFM imaging of a single Aspergillus fumigatus. (a) AFM height image of a single cell trapped in polymeric membrane. (b) Deflection AFM image showing the crystalline rodlet layers. (c) High-resolution imaging of the squared area in (b) of the rodlets of hydrophobins. (d) AFM cross-section (line in (c)), showing hudrophobins height. Reprinted from ref. [17] with permission from Elsevier.

dynamic structure that continuously evolves from conidial formation to germination and play a role during the early establishment of aspergillosis. Progress in understanding the mechanisms of self-assembly and the rodlet structure may also lead to advances toward solving medical problems associated with amyloid aggregation [19].

Similarly, great efforts have been made to visualize S-layer nanoarrays on living *Corynebacterium glutamicum* bacterial cells [6]. The first direct AFM observation of crystalline S-layers on live bacteria showed hexagonal unit cells with dimensions similar to those reported for isolated S-layer sheets. Most interestingly, AFM-live cell imaging also revealed a new inner layer composed of periodic nanogrooves, presumably reflecting the unique specificity of the *C. glutamicum* cell wall (Figure 3.3) [20]. It was suggested that this second nanostructured layer could function as a biomolecular template promoting the two-dimensional assembly and crystallization of S-layer monomers [21]. The in vivo visualization of S-layer has allowed a better understanding of the structure of the protein monomolecular array in its native state [6, 21].

One of the major advances in using AFM in microbiology for the last decade is the revelation of the three-dimensional structure of peptidoglycan which is the main constituent of bacterial cell walls. Despite the important functional roles of this polymer (mechanical strength, cell shape, and target for antibiotics), its three-dimensional organization has long been controversial [22–24]. In the most widely accepted model, glycan strands run parallel to the plasma membrane, arranged perhaps as hoops or helices around the short axis of the cell, resulting in a woven fabric. In past years, AFM imaging has complemented electron cryo-microscopy and

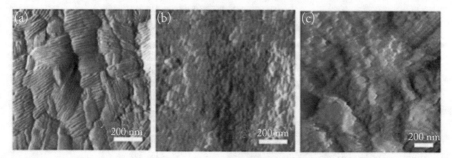

Figure 3.2: Nanoscale imaging of live cells. (a) AFM deflection image in aqueous solution revealing rodlets on the surface of an *Aspergillus fumigatus* spore. (b) Image recorded on the same spore after 3 h germination, indicating that the crystalline rodlet layer has disappeared, revealing the underlying amorphous polysaccharide cell wall. (c) Image obtained for an *A. fumigatus* mutant deficient in rodlet production showing similar morphology as in (b). Reprinted with permission from ref. [20]. Copyright (2009) American Chemical Society.

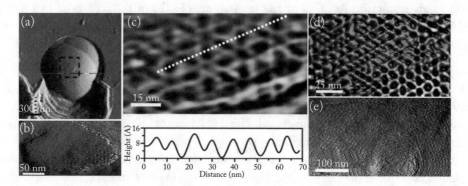

Figure 3.3: (a) AFM deflection image of a *C. glutamicum* cell grown in a heart-brain infusion show patches of ordered S layers. The image in (b) corresponds to the square region shown in (a). (c) AFM cross-section in taken along the dotted lines. (d) AFM image showing a well-ordered S-layer. (e) AFM reveals a new inner layer of highly ordered material. Reprinted with permission from ref. [20]. Copyright (2009) American Chemical Society.

tomography techniques in providing key structural details of peptidoglycan, such as strand orientation. The organization of peptidoglycan has been first visualized in living cells, as shown in Figure 3.4a, AFM high-resolution imaging of the pathogen *Staphylococcus aureus* in growth medium showed ring-like and honeycomb structures attributed to peptidoglycan [7, 22]. Similarly, 25-nm-wide periodic bands running parallel to the short axis of the cell were found on *Lactococcus lactis* mutant strains lacking cell wall exopolysaccharides feature (Figure 3.4b) [24]. Also, AFM high-resolution images of *Bacillus atrophaeus* spores during germination revealed a

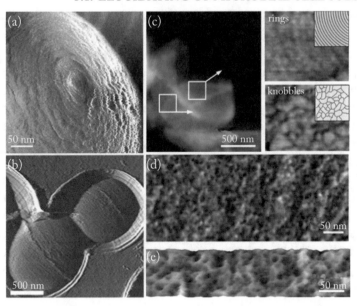

Figure 3.4: Unravelling peptidoglycan architecture. (a) Deflection AFM image of a newly formed intact *S. aureus* cell wall, showing concentric rings that surround a central depression. (b) Deflection AFM image of dividing bacteria recorded for *L. lactis* VES5748 WPS—mutant lacking cell wall exopolysaccharides. Reprinted from ref. [24] with permission from Springer Nature. Copyright (2010). (c–e) Deflection AFM images of purified sacculi from *E. coli* (c), emphasizing key architectural details: concentric rings and knobbly surface structures, and (d, e) bands of porous material running circumferentially around the sacculi. Reprinted with permission from ref. [6]. Copyright (2014) American Chemical Society.

porous network of peptidoglycan fibers, consistent with a honeycomb model structure for synthetic peptidoglycan oligomers [25]. However, much of the AFM/peptidoglycan studies have been carried out on purified sacculi by the Foster research team [22, 26–29]. In an initial study, they reported that the cell wall of the model rod-shaped bacterium *Bacillus subtilis* has glycan strands up to 5 μm, thus longer than the cell itself [8]. The inner surface of the cell wall showed 50-nm-wide peptidoglycan cables running parallel to the short axis of the cell, together with cross striations with an average periodicity of 25 nm along each cable. The data favored an architectural model where glycan strands are polymerized and cross-linked to form a peptidoglycan rope, which is then coiled into a helix to form the inner surface cable structures [22]. In another study, AFM was combined with optical microscopy with fluorescent vancomycin labeling to investigate the distribution of peptidoglycan in the spherical bacterium *Staphylococcus aureus* [27]. Concentric rings and knobbly surface structures were observed and attributed to nascent and mature peptidoglycan, respectively (Figure 3.4c–e). Peptidoglycan features were suggested to demark previous divisions and, in doing so, hold the necessary information to specify the next division plane. Peptidoglycan architecture and dynamics have also been investigated in bacteria with ovoid cell shape (ovococci), including a number of important pathogens [26]. In these species, AFM images showed a preferential orientation of the peptidoglycan network parallel to the short axis of the cells, while superresolution fluorescence microscopy unraveled the dynamics of peptidoglycan assembly. The results suggested that ovococci have a unique peptidoglycan architecture not observed previously in other model organisms. Recently, the rod-shaped Gram-negative bacterium *Escherichia coli* was shown to feature peptidoglycan structures running parallel to the plane of the sacculus but in many directions relative to the long axis [29]. The images also revealed bands of porosity running circumferentially around the sacculi (Figure 3.4d,e). Super-resolution fluorescence microscopy unraveled an unexpected discontinuous, patchy synthesis pattern. A model was suggested in which only the more porous regions of the peptidoglycan network are permissive for synthesis.

Accordingly, these high-resolution studies have shown that bacterial species exhibit a variety of peptidoglycan architectures, thereby contributing to new structural models of peptidoglycan arrangement [22].

3.2   IMAGING DYNAMIC PROCESSES ON LIVING BACTERIA

The possibility to image single live cells in real-time provides a novel insight into the dynamics of the cell surface. The technique has allowed the monitoring of bacterial growth and division under physiological conditions of several bacteria species [6, 7, 15, 17, 30–36]. In one of the first AFM elegant study, cell growth, and division events in *S. aureus* were monitored using AFM combined with thin-section transmission electron microscopy [7]. Nanoscale holes were seen around the septal annulus at the onset of division and attributed to cell wall structures possessing high autolytic activity (Figure 3.5a–h) [7, 35]. After cell separation, concentric rings were observed on the surface of the new cell wall and suggested to reflect newly formed peptidoglycan (Figure 3.4a). This data shows that immobilized organisms, using the pore trapping methodology, are sufficiently well trapped to observe dynamic processes with near molecular resolution. Also, AFM can provide valuable pieces of information on growth and division processes, which are complementary to the data obtained using other microscopies. Using the AFM dynamic jumping mode, Van Der Hofstadt et al. have been able to image living single bacterial cells belonging to two different *Escherichia coli* strains, the MG1655 and the enteroaggregative (EAEC) 042, both being weakly adsorbed onto planar gelatin coated substrates [30]. In addition, the authors have been able to monitor the growth and division of *E. coli* 042 in its native state over long periods of time (Figure 3.5i–n). This method has made it possible to observe the bacterial growth and division, an event which is difficult to image in real time for gram-negative species because of the flagella and pili around the cell. These results open new possibilities in the in-situ observation of living bacterial processes at the single cell and nanoscale levels. Similarly, AFM contact

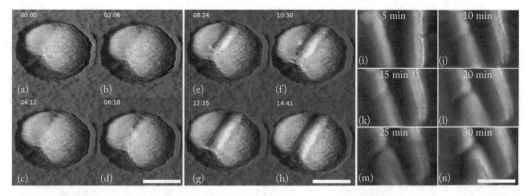

Figure 3.5: (a–h) AFM phase images of division of *S. aureus* (NCTC 8532) under brain heart infusion (BHI) at room temperature (times displayed as minutes: seconds from first observable signs of division): (a, b) first signs of division; (c) further invagination of cell wall; (d–g) the outer part of the cell wall breaks; and (h) cells have divided. Reprinted from ref. [35] with permission from Wiley. (i–n) Time sequence AFM deflection images of growing E. coli 042 bacteria in its natural growing medium (DMEM + 0.45% glucose) for its aggregation growth on gelatinized coatings of mica. The formation of the septum in ∼ 15 min can be observed on the left bacteria. Reprinted from ref. [30] with permission from Elsevier. Scale bar: 1 μm.

mode imaging was used to investigate the morphological and nanostructural surface changes in *E. coli* over time [34]. These bacteria were found to collapse and release some of their cellular components after a week, which became, in turn, a nutrient source for weakened bacteria. Most importantly, along with the morphological evolution of bacteria, their outer membrane exhibits strong signs of re-organization with time: "ripples," characteristic of healthy bacteria, turn into spherical aggregates with irregular boundaries. These modifications are likely due to the reorganization of LPS molecules, major constituents of Gram-negative bacteria external membrane, under the exposure of external stress [34]. In addition, to bacteria growth and division investigations, interesting studies were performed on spores and other microbial cells. A good example of these studies is the germination of *A. fumigatus conidia* [16], during which progressive disruption of the rodlet layer revealed the underlying inner cell wall structures. High-resolution AFM imaging was used to investigate the dynamic of the assembly of the rodlet layer on the surface of the conidium over time [19]. The appearance and the number of the rodlets on the surface of the conidium was dependent on the conidial age. The maximal rodlet formation was seen after 30 days of culture without further evolution of the rodlet morphology afterward.

In addition to the real-time investigations of bacterial growth and division processes and owing to its ability to monitor drug-induced surface alterations in microbial pathogens, AFM has opened up new possibilities for understanding the mode of action of antibiotics and for screening new antimicrobial molecules capable to fight resistant strains [14, 37–41]. For ex-

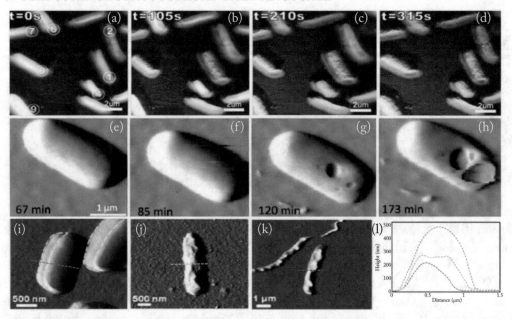

Figure 3.6: (a–d) AFM images of early stage kinetics of CM15 action on bacteria. Images recorded every 105 sec. Reprinted from ref. [38] with permission from Springer Nature. Copyright (2010); (e–h) AFM deflection images of wild-type *K. pneumoniae* cells after addition of 6 μM caerin peptide solution. Pore forming peptide action is shown after 120 min exposure time. Reprinted from ref. [42] with permission from Elsevier. (i–l) AFM deflection images recorded on (i) native, (j) ticarcillin-, and (k) tobramycin-treated *P. aeruginosa* cells; (l) Vertical AFM cross-sections taken along the dashed lines of native (blue), ticarcillin- (green), and tobramycin-treated cell (red). Reprinted from ref. [43] with permission from Springer Nature. Copyright (2012).

ample, Fantner et al. investigate the real-time kinetics of single *Escherichia coli* cells death using high-speed AFM38. As shown in Figure 3.6a–d, the combination of very soft but rapid imaging (13 sec per image) allowed to monitor multiple cells at once and the characterization of the initial stages of the action of the antimicrobial peptide CM15 on individual cells with nanometer resolution. The most obvious effect of the addition of CM15 is a change in the surface state of the bacteria from smooth to corrugated (Figure 3.6a–d). Interestingly, there is a wide range in the time of onset of the change for individual bacteria. Bacterium 1 in Figure 3.6a starts changing within 13 sec of the addition of CM15, and the change is completed in $\sim$ 60 sec. Bacterium 2 does not start changing until $\sim$ 80 sec, and the change is not complete until $\sim$ 120 sec. These results suggest that the killing process is a combination of a time-variable incubation phase (which takes seconds to minutes to complete) and a more rapid execution phase [38].

In a recent study, Mularski et al. used time-resolved AFM imaging to monitor the effect of the antimicrobial peptide caerin 1.1 on *lysed Klebsiella pneumoniae* cells, corroborating a pore-forming mechanism of action [42]. Figure 3.6e–h, shows the time course of the formation of depressions (holes) on the bacterial surface caused by the antimicrobial peptide. Also, the data of this study demonstrated that the presence of a capsule confers no advantage to wild-type over capsule-deficient cells when exposed to the caerin. In addition to AFM real-time imaging the authors performed AFM force pushing on bacteria cells to correlate the bacteria turgor pressure with the structural effects due to the antimicrobial peptide. Caerin 1.1 appears to affect the mechanobiology of bacteria via membrane pore formation. In a similar study, *Pseudomonas aeruginosa* cell wall was demonstrated to be biophysically affected at the nanoscale by two reference antibiotics, ticarcillin, and tobramycin, with the elasticity dropping dramatically after treatment [40]. AFM high-resolution images were recorded on single bacteria before and after treatment to qualitatively explore the impact of tobramycin and ticarcillin on cell surface. As shown in Figure 3.6i–l, the native cell shows a smooth surface while the morphology of the treated cells was completely lost. In a related study, the effect of a polycationic calixarene-based guanidinium compound, CX1, on an *P. aeruginosa* multidrug-resistant strain was investigated by the same research group [43]. CX1 caused substantial alteration of the cell wall morphology (increased roughness and perforations) and a major drop in the cell wall stiffness. Using real-time imaging, Francius et al. [44] captured the structural dynamics of *S. aureus* cells exposed to lysostaphin, an enzyme that specifically cleaves the peptidoglycan cross-linking pentaglycine bridges and that represents an interesting potential alternative to antibiotics. The enzyme induced major changes in cell surface morphology (swelling, splitting of the septum, and nanoscale perforations) and cell wall mechanics, which were attributed to the digestion of peptidoglycan, leading eventually to the formation of osmotically fragile cells [44].

In summary, these measurements demonstrate the enormous potential of AFM and its modalities for assessing and monitoring dynamic processes in microbial cells with high resolution and at the single cell level. The advances in AFM technology and its application reported here constitute an enabling technology with which cell biologists may explore cellular processes, in real time, at the nanometer level and under various conditions.

3.3   REFERENCES

[1] Janakiraman, A. and Goldberg, M. B. Recent advances on the development of bacterial poles. *Trends Microbiol.*, 12:518–525, 2004. DOI: 10.1016/j.tim.2004.09.003. 39

[2] Ebersbach, G. and Jacobs-Wagner, C. Exploration into the spatial and temporal mechanisms of bacterial polarity. *Trends Microbiol.*, 15:101–108, 2007. DOI: 10.1016/j.tim.2007.01.004.

[3] Maddock, J. R., Alley, M. R. K., and Shapiro, L. Minireview polarized cells, polar actions. *J. Bacteriol.*, 175(22):7125–7129, 1993. DOI: 10.1128/jb.175.22.7125-7129.1993.

[4] Brehm-Stecher, B. F. and Johnson, E. A. Single-cell microbiology: Tools, technologies, and applications. *Microbiol. Mol. Biol. Rev.*, 68:538–59, 2004. DOI: 10.1128/mmbr.68.3.538-559.2004. 39

[5] Lidstrom, M. E. and Konopka, M. C. The role of physiological heterogeneity in microbial population behavior. *Nat. Chem. Biol.*, 6:705–712, 2010. DOI: 10.1038/nchembio.436. 39

[6] Dufrêne, Y. F. Atomic force microscopy in microbiology: New structural and functional insights into the microbial cell surface. *mBio*, 5(4):e01363–14, 2014. DOI: 10.1128/mbio.01363-14. 40, 41, 43, 44

[7] Touhami, A., Jericho, M. H., and Beveridge, T. J. Atomic force microscopy of cell growth and division in Staphylococcus aureus. *J. Bacteriol.*, 186:3286–95, 2004. DOI: 10.1128/jb.186.11.3286-3295.2004. 40, 42, 44

[8] Turner, R. D., Thomson, N. H., Kirkham, J., and Devine, D. Improvement of the pore trapping method to immobilize vital coccoid bacteria for high-resolution AFM: A study of staphylococcus aureus. *J. Microsc.*, 238:102–110, 2010. DOI: 10.1111/j.1365-2818.2009.03333.x. 40, 43

[9] Plomp, M., Leighton, T. J., Wheeler, K. E., Hill, H. D., and Malkin, A. J. In vitro high-resolution structural dynamics of single germinating bacterial spores. *Proc. of the Natl. Acad. Sci.*, 104, National Academy of Sciences, 2007. DOI: 10.1073/pnas.0610626104. 40

[10] Alex, G., Larry, W., and Yun, X. Nanomechanical characterization of bacillus anthracis spores using atomic force microscopy. *Appl. Environ. Microbiol.*, 82, 2016. DOI: 10.1128/aem.00431-16. 40

[11] Van Der Hofstadt, M., Hüttener, M., Juárez, A., and Gomila, G. Nanoscale imaging of the growth and division of bacterial cells on planar substrates with the atomic force microscope. *Ultramicroscopy*, 154:29–36, 2015. DOI: 10.1016/j.ultramic.2015.02.018. 40

[12] Touhami, A., Jericho, M. H., Boyd, J. M., and Beveridge, T. J. Nanoscale characterization and determination of adhesion forces of Pseudomonas aeruginosa pili by using atomic force microscopy. *J. Bacteriol.*, 188:370–377, 2006. DOI: 10.1128/jb.188.2.370-377.2006. 40

[13] Stukalov, O., Korenevsky, A., Beveridge, T. J., and Dutcher, J. R. Use of atomic force microscopy and transmission electron microscopy for correlative studies of bacterial capsules. *Appl. Environ. Microbiol.*, 74:5457–5465, 2008. DOI: 10.1128/aem.02075-07. 40

[14] Verbelen, C. et al. Ethambutol-induced alterations in mycobacterium bovis BCG imaged by atomic force microscopy. *FEMS Microbiol. Lett.*, 264:192–197, 2006. DOI: 10.1111/j.1574-6968.2006.00443.x. 40, 45

[15] Tripathi, P. et al. Towards a nanoscale view of lactic acid bacteria. *Micron*, 43:1323–1330, 2012. DOI: 10.1016/j.micron.2012.01.001. 40, 44

[16] Dague, E., Alsteens, D., Latgé, J.-P., and Dufrêne, Y. F. High-resolution cell surface dynamics of germinating aspergillus fumigatus Conidia. *Biophys. J.*, 94:656–660, 2008. DOI: 10.1529/biophysj.107.116491. 40, 45

[17] Alsteens, D. *Microbial Cells Analysis by Atomic Force Microscopy. Methods in Enzymology*, 506, Elsevier Inc., 2012. DOI: 10.1016/b978-0-12-391856-7.00025-1. 40, 41, 44

[18] Alsteens, D. et al. Nanomicrobiology. *Nanoscale Res. Lett.*, 2:365–372, 2007. DOI: 10.1007/s11671-007-9077-1. 40

[19] Valsecchi, I. et al. Assembly and disassembly of aspergillus fumigatus conidial rodlets. *Cell Surf.*, 5:100023, 2019. DOI: 10.1016/j.tcsw.2019.100023. 41, 45

[20] Dupres, V., Alsteens, D., Pauwels, K., and Dufrêne, Y. F. In vivo imaging of s-layer nanoarrays on covynebactevium glutamicum. *Langmuir*, 25:9653–9655, 2009. DOI: 10.1021/la902238q. 41, 42

[21] Alsteens, D., Dupres, V., Andre, G., and Dufrêne, Y. F. Frontiers in microbial nanoscopy. *Nanomedicine*, 6:395–403, 2011. DOI: 10.2217/nnm.10.151. 41

[22] Turner, R. D., Vollmer, W., and Foster, S. J. Different walls for rods and balls: The diversity of peptidoglycan. *Mol. Microbiol.*, 91:862–874, 2014. DOI: 10.1111/mmi.12513. 41, 42, 43, 44

[23] Vollmer, W. and Seligman, S. J. Architecture of peptidoglycan: More data and more models. *Trends Microbiol.*, 18:59–66, 2010. DOI: 10.1016/j.tim.2009.12.004.

[24] Andre, G. et al. Imaging the nanoscale organization of peptidoglycan in living Lactococcus lactis cells. *Nat. Commun.*, 1:1–8, 2010. DOI: 10.1038/ncomms1027. 41, 42, 43

[25] Plomp, M. et al. Spore coat architecture of clostridium novyi NT spores. *J. Bacteriol.*, 189:6457–6468, 2007. DOI: 10.1128/jb.00757-07. 43

[26] Wheeler, R., Mesnage, S., Boneca, I. G., Hobbs, J. K., and Foster, S. J. Super-resolution microscopy reveals cell wall dynamics and peptidoglycan architecture in ovococcal bacteria. *Mol. Microbiol.*, 82:1096–1109, 2011. DOI: 10.1111/j.1365-2958.2011.07871.x. 43, 44

[27] Turner, R. D. et al. Peptidoglycan architecture can specify division planes in staphylococcus aureus. *Nat. Commun.*, 1:1–9, 2010. DOI: 10.1038/ncomms1025. 43

[28] Hayhurst, E. J., Kailas, L., Hobbs, J. K., and Foster, S. J. Cell wall peptidoglycan architecture in bacillus subtilis. *Proc. Natl. Acad. Sci.*, 105:14603–14608, 2008. DOI: 10.1073/pnas.0804138105.

[29] Turner, R. D., Hurd, A. F., Cadby, A., Hobbs, J. K., and Foster, S. J. Cell wall elongation mode in gram-negative bacteria is determined by peptidoglycan architecture. *Nat. Commun.*, 4:1496–1498, 2013. DOI: 10.1038/ncomms2503. 43, 44

[30] Van Der Hofstadt, M., Hüttener, M., Juárez, A., and Gomila, G. Nanoscale imaging of the growth and division of bacterial cells on planar substrates with the atomic force microscope. *Ultramicroscopy*, 154:29–36, 2015. DOI: 10.1016/j.ultramic.2015.02.018. 44, 45

[31] Plomp, M., Carroll, A. M., Setlow, P., and Malkin, A. J. Architecture and assembly of the bacillus subtilis spore coat. *PLoS One*, 9:e108560, 2014. DOI: 10.1371/journal.pone.0108560.

[32] Yamashita, H. et al. Single-molecule imaging on living bacterial cell surface by high-speed AFM. *J. Mol. Biol.*, 422:300–309, 2012. DOI: 10.1016/j.jmb.2012.05.018.

[33] Alsteens, D. et al. Nanoscale imaging of microbial pathogens using atomic force microscopy. *Wiley Interdiscip. Rev. Nanomedicine Nanobiotechnology*, 1:168–180, 2009. DOI: 10.1002/wnan.18.

[34] Gammoudi, I. et al. Morphological and nanostructural surface changes in escherichia coli over time, monitored by atomic force microscopy. *Colloids Surf. B Biointerf.*, 141:355–364, 2016. DOI: 10.1016/j.colsurfb.2016.02.006. 45

[35] Turner, R. D., Thomson, N. H., Kirkham, J., and Devine, D. Improvement of the pore trapping method to immobilize vital coccoid bacteria for high-resolution AFM: A study of staphylococcus aureus. *J. Microsc.*, 238:102–110, 2010. DOI: 10.1111/j.1365-2818.2009.03333.x. 44, 45

[36] Chen, P. et al. Nanoscale probing the kinetics of oriented bacterial cell growth using atomic force microscopy. *Small*, 10:3018–3025, 2014. DOI: 10.1002/smll.201303724. 44

[37] Quilès, F., Saadi, S., Francius, G., Bacharouche, J., and Humbert, F. In situ and real time investigation of the evolution of a pseudomonas fluorescens nascent biofilm in the presence of an antimicrobial peptide. *Biochim. Biophys. Acta—Biomembr.*, 1858:75–84, 2016. DOI: 10.1016/j.bbamem.2015.10.015. 45

[38] Fantner, G. E., Barbero, R. J., Gray, D. S., and Belcher, A. M. Kinetics of antimicrobial peptide activity measured on individual bacterial cells using high-speed atomic force microscopy. *Nat. Nanotech.*, 5:280–285, 2010. DOI: 10.1038/nnano.2010.29. 46

[39] Longo, G. et al. Antibiotic-induced modifications of the stiffness of bacterial membranes. *J. Microbiol. Meth.*, 93:80–84, 2013. DOI: 10.1016/j.mimet.2013.01.022.

[40] Formosa, C., Grare, M., Duval, R. E., and Dague, E. Nanoscale effects of antibiotics on pseudomonas aeruginosa. *Nanomed. Nanotechnol., Biol. Med.*, 8:12–16, 2012. DOI: 10.1016/j.nano.2011.09.009. 47

[41] Pinzó N-Arango, P. A., Scholl, G., Nagarajan, R., Mello, C. M., and Camesano, T. A. Atomic force microscopy study of germination and killing of bacillus atrophaeus spores. *J. Molec. Recog.*, 2009. DOI: 10.1002/jmr.945. 45

[42] Mularski, A., Wilksch, J. J., Hanssen, E., Strugnell, R. A., and Separovic, F. Atomic force microscopy of bacteria reveals the mechanobiology of pore forming peptide action. *Biochim. Biophys. Acta—Biomembr.*, 1858:1091–1098, 2016. DOI: 10.1016/j.bbamem.2016.03.002. 46, 47

[43] Formosa, C. et al. Nanoscale analysis of the effects of antibiotics and CX1 on a pseudomonas aeruginosa multidrug-resistant strain. *Sci. Rep.*, 2:575, 2012. DOI: 10.1038/srep00575. 46, 47

[44] Francius, G., Domenech, O., Mingeot-Leclercq, M. P., and Dufrêne, Y. F. Direct observation of staphylococcus aureus cell wall digestion by lysostaphin. *J. Bacteriol.*, 190:7904–7909, 2008. DOI: 10.1128/jb.01116-08. 47

C H A P T E R 4

AFM Force Spectroscopy of Living Bacteria

OVERVIEW

Bacterial surfaces rely on a complex set of multicomponent cellular structures to perform vital functions such as, cell growth and division, adhesion, and aggregation. These structures often adopt specific shapes, reside in specific subcellular spaces, and carry out functions specific to these properties. For example, the macromolecular cell wall structure forms a barrier to separate the inner cell from its environment and fulfills essential functions such as defining cell shape, cell motility, and cell adhesion processes [1–3]. Understanding how these cellular structures and machineries are assembled at the correct time and space to achieve their functions are important not only in microbiology, to elucidate cellular functions (such as ligand-binding or biofilm formation), but also in medicine (biofilm infections) [4] and biotechnology (cell aggregation) [5]. AFM multiparametric imaging, using either bare tips or tips functionalized with chemical groups, ligands, or cells, has recently offered exciting new possibilities to quantitatively map the biophysical properties of bacterial surfaces under physiological conditions [6, 7]. FD curve-based AFM in particular has enabled researchers to map and quantify biophysical properties and biomolecular interactions of a wide variety of bacterial species. Rapid progress in FD-based multiparametric imaging modalities have been developed, allowing us to simultaneously image the structure, elasticity, and interactions of biological samples at high spatiotemporal resolution.

By oscillating the AFM tip, spatially resolved FD curves are obtained at much higher frequency than before, and as a result, samples are mapped at a speed similar to that of conventional topographic imaging [8]. In addition, the combination of two or more AFM-based modalities to characterize multiple parameters of complex bacterial surfaces, of which FD-based AFM is a prominent example, increases the diversity and volume of data that can be acquired in an experiment [9]. Such combination allows, for example, correlation of ligand-binding events to topographies of living cells. Forces driving cell-cell and cell-substrate interactions can be now recorded simultaneously with high-resolution imaging of dynamic processes, leading to better understanding of the forces driving cell adhesion and biofilm formation [7, 9].

In this chapter, we discuss the general principle of AFM force spectroscopy and multiparametric AFM imaging and we provide a snapshot of recent studies showing how this new technology has been applied to investigate bacterial surfaces. We emphasize methodologies in which multiparametric imaging is combined with probes functionalized with chemical groups, ligands, or even live cells, in order to image and quantify receptor interaction forces and free-energy landscapes in a way not possible before. This includes imaging the sites of assembly and extrusion of single filamentous bacteriophages in living bacteria, unravelling the adhesive properties of biofilm-forming microbial pathogens, mapping cell-cell and cell-solid interactions and the label-free detection of bioanalytes and cells. In the coming years, it is anticipated that multiparametric AFM imaging will be increasingly used by scientists from broad horizons, enabling them to shed light into the sophisticated functions of biomolecular and cellular systems.

4.1   BASICS OF AFM FORCE SPECTROSCOPY

Although AFM was originally invented for structural imaging the topography of surfaces, it has recently evolved into a multifunctional molecular toolbox, enabling researchers to probe biophysical properties of biological systems over scales ranging from single molecules to whole cells [7, 10]. Thus, the relative ease of use and the commercial availability of the AFM make it one of the most widely used instruments for measuring molecular forces. Three main force spectroscopy modalities are often used to probe these properties at the surface of living bacteria. A single molecule force spectroscopy (SMFS) is the most widely used mode and it consists in measuring the interactions between bear or modified AFM tips and single molecules (protein, polymer, pili, flagella), at the cell surface (Figure 4.1a) [11–15]. The molecular recognition mapping (MRM) mode was developed for mapping specific interaction between two specific molecules simultaneously with topography imaging (Figure 4.1b) [7, 15–18]. The specific interaction occurs between a molecule attached to the AFM tip (chemical modification or immobilized ligand) and a molecule at the bacteria surface. Topography and molecular recognition images can be compared to determine correlation of binding events and topographic profiles [15, 18]. As shown in Figure 4.1c, force spectroscopy may also be applied to whole cells, called single-cell force spectroscopy (SCFS), offering a means to study heterogeneity and interactions in cellular communities [9, 13]. In SMFS and MRM, the tip is functionalized with biological molecules, such as proteins, carbohydrates, or even viruses, while in SCFS the tip is replaced by a living cell. SCFS is frequently combined with optical microscopy to monitor cell morphology during an adhesion experiment or to localize fluorescently labeled proteins. The cell morphology changing in response to mechanical forces can be well monitored by confocal or spinning disk confocal microscopy [1, 19].

As detailed in Chapter 1, AFM force spectroscopy is performed by approaching the AFM tip to and retracting it from the biological sample while recording single FD curves. The approach curve (blue curve in Figure 4.1d) can be used to quantify the height, electrostatic charge, mechanical deformation, and elastic modulus of the sample [20]. The unbinding events recorded during tip (or cell) retraction (red curve in Figure 4.1d) is the parameter that provides key information on the adhesion and mechanical properties of individual biomolecules in SMFS, receptor distribution in MRM, and single-cell adhesion forces in SCFS [7, 11, 21]. To reliably charac-

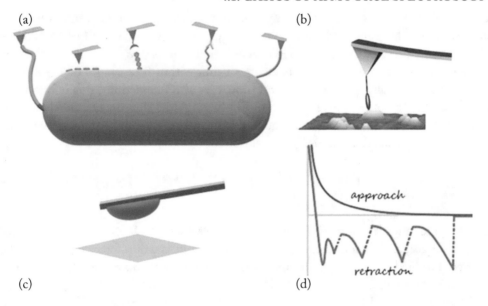

Figure 4.1: (a) An illustration of single-molecule force spectroscopy (SMFS) for probing the localization, adhesion, and interactions of individual surface molecules and appendages on a single bacterium. (b) An illustration of MRM: an AFM tip-tethered antibody binds to its antigen in the sample being scanned. (c) Single-cell force spectroscopy principle. (d) An example of force-distance (FD) curve showing the approach and retraction parts.

terize the properties of the sample implies precisely controlling the interaction between tip and sample, thus requiring AFM tips with well-defined geometry and surface chemistry [7, 17]. Sophisticated commercial micro-and nano-machined cantilevers and tips are now available, which are customized in terms of shape, tip radius, and physical and chemical properties. Two interconnected issues in FD-based AFM are the lateral and temporal resolutions [8]. In modern AFMs, the lateral resolution is mainly related to the tip radius, the tip-sample drift, the distance dependence of the tip-sample interaction, imaging force, and the properties of the biological sample. Long-range surface forces interacting over several tens of nanometers reduce the resolution at which these interactions can be localized [20, 22]. Technically, when recording an AFM image at a certain frame size, the number of pixels recorded determines the theoretically approachable resolution. However, the number of pixels and thus the amount of force curves collected per FD-based AFM image is limited by the data acquisition time [6]. In the early days of FD-based AFM, the time required for recording a single force curve was between ~ 0.1 and 10 sec, and the time needed to acquire a FD-based AFM image of 32 pixels $\times$ 32 pixels ranged from $\sim$ 2 min to ~ 3 h [6]. This slow imaging speed strongly limited the use of FD-based AFM imaging in biology, but the recent introduction of faster piezo elements, feedback loops, data acquisition systems, oscillation modes changing the tip-sample distance, and tailored cantilevers reducing hydrodynamic drag largely solved this problem. As a consequence, modern FD-based AFMs can record several hundreds of thousands of FD curves simultaneously with multiparametric images of native biosystems at a resolution approaching 1 nm, within time ranges of 15–30 min [6]. As each FD curve locally quantifies physical properties and interactions, this information can be directly mapped to the sample topography. FD-based AFM thus opened the door to imaging complex biological systems and to simultaneously quantifying and mapping their intrinsic physical properties, including elasticity and adhesion.

4.2   SINGLE-MOLECULE FORCE SPECTROSCOPY (SMFS)

Often SMFS requires labeling the AFM tip with molecules that interact with specific targets on the cell surface, and then measuring the characteristic interaction forces between the ligand and its receptor [15]. A variety of biofunctionalization protocols have been developed for this purpose with the two most common methods being silanization (Figure 4.2a), and exploitation of simple thiol-gold bonding on gold-coated surfaces (Figure 4.2b) [23]. A bifunctional crosslinker molecules such polyethylene glycol (PEG) can be used to anchor ligand firmly at low density, while maintaining its conformational flexibility, and functionality [6, 7, 23]. However, self-assembled monolayers of thiols on gold surfaces are most preferred for biofunctionalization of AFM tips for SMFS studies [9, 11, 18, 24, 25]. Briefly, a thin layer of gold (50–100 nm) can be sputtered onto a standard silicon AFM probe. The sensing element can then be directly bound to the AFM tip via suitable thiol, as shown in Figure 4.2b. Also, a bifunctional linker such as PEG with at least one thiol end can be used to attach the ligand to the gold-AFM tip. Again, a long PEG linker reduces non-specific tip-sample interactions and allows the bound recognition element the conformational flexibility needed bind its cognate ligand. SMFS may also be exploited to pull on single molecules or polymeric structures at the cell surface, such as proteins, pili, and flagella (Figure 4.1a), in order to investigate their mechanical properties that play an important role in cell behavior. These single-molecule manipulations have shown that microbial molecules and structures feature unanticipated mechanical responses when subjected to force (for example, protein unfolding and unzipping, pilus extension, and spring properties), which contribute significantly to cellular functions such as mechanosensing and adhesion [8].

In a standard SMFS experiment, thousands of FD curves are collected from different locations on the cell surface, using a bear or functionalized AFM tip. The shape and features in the FD curve reflect the type of interactions between the tip and the cell surface. For example, a force curve as shown in Figure 4.2c is generally attributed to AFM tip stretching polymeric structure such as bacterial pili or polysaccharides [22]. FD curves obtained by pulling on proteins have a characteristic sawtooth pattern corresponding to the sequential unfolding of the molecule as it is stretched and the bonds holding the molecule or domains together break down. As shown in Figure 4.2d, the sawtooth peaks on the force curves can correspond to the unfold-

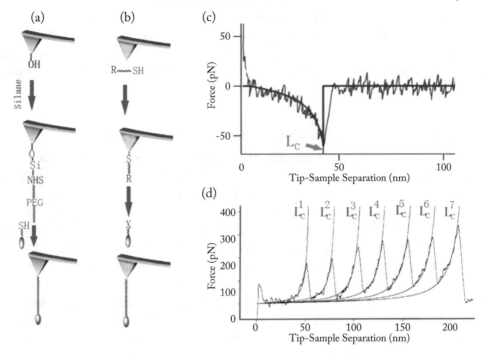

Figure 4.2: (a–b) Main steps of the two functionalization methods for AFM tips: (a) silanization of a silicon tip and (b) thiolation of gold-coated tip. (c) an example of FD curve showing a single adhesion peak and the worm-like-chain (WLC) fit (red curve). (d) An example of FD curve for protein-domain unfolding by AFM force pulling. The distance between two contour lengths (L_c) corresponds to the protein domain length.

ing of subparts of the protein such as β-strands, α-helices, globular domains, or entire protein molecules [7, 20, 21]. Often, analyses of AFM force curves are carried out using models such as freely jointed chain (FJC) [26] or wormlike chain model (WLC), which is commonly applied and is accurate up to stretching forces of several hundred piconewtons. Mathematically, the WLC is described by the equation 1, that relates the stretching force to the persistence length and the contour length of the stretched molecule [26]:

$$F(x) = \frac{k_B T}{P} \left[\frac{1}{4} \left(1 - \frac{x}{L_c} \right)^{-2} + \frac{x}{L_c} - \frac{1}{4} \right], \tag{4.1}$$

where L_c is the contour length of the polymer (polypeptide), x is the polymer extension, and P is the persistence length of the polymer, which describes the rigidity of the polymer and the distance over which the chain orientation is lost [26, 27]. Thus, important information about these interaction forces can be deduced (contour length, persistence length, rupture length, and rupture force) by performing statistical analysis and curve fitting using polymer models [7]. Because the protein behaves approximately as an elastic polymer chain in between unfolding events, WLC can be fitted to the peaks in the force curves to obtain information about the contour length released by each unfolding event and the persistence length of the unfolded segment (Figure 4.2d) [28].

4.2.1   PULLING ON PROTEINS ON LIVE BACTERIA CELLS

Probing the spatial distribution, biding affinity, and mechanical strength of cell envelope constituents is a crucial task in current cellular microbiology [11]. During the last two decades, many studies have shown that SMFS can localize individual proteins and polysaccharides on the surface of live cells, and measure their specific binding strength [7, 11, 16, 25, 29, 30]. Nowadays, SMFS experiments are more frequently used to determine how cell envelope constituents are organized in space and to probe their mechanical properties. As result, there is increasing evidence that cell membranes and cell walls display heterogeneous organizations [11]. For example, early SMFS study by Dupres et al. showed that heparin-binding haemagglutinin adhesins produced by *Mycobacterium tuberculosis*, engaged in host interactions, were concentrated into nanodomains to promote the recruitment of receptors in host cells [15]. This force-induced clustering of adhesins mechanism, for activating cell adhesion in microbial pathogens, was also demonstrated in *Candida. albicans* using AFM tips terminated with specific antibodies to probe the agglutinin-like sequence (*Als*) adhesins on living cells [31]. In a recent study, Hinterdorfer and co-workers, exploited SMFS to reveal essential details about how bacterial curli fibers binds to fibronectin (FN) [32]. The curli fibers are produced by many enteric bacteria, including pathogenic *Escherichia coli* and *Salmonella* strains to facilitate their attachment to host surfaces, leading to bacterial internalization into host cells [33–36]. In particular, the study focused on comparing binding forces of various fibronectin constructs to curli both in its monomeric isolated form and in its oligomeric fiber state expressed on bacterial surfaces. As shown in Figure 4.3a, AFM cantilever tips were functionalized with multi-domain full-length fibronectin (FN), isolated domain III (FN III), or a peptide with the core RGD sequence (RGD). Dimeric FN consists of two FNIII domains, each of which contains one wide-spread and specific binding sequence RGD. The functionalized tips were repeatedly brought into contact with living *E. coli* bacteria overexpressing the curli protein CsgA (CsgA$^+$ mutant) and CsgA knock-out mutant (CsgA$^-$). Compare to the CsgA$^+$ mutant, the mutant lacking curli expression showed a smoother surface structure (Figure 4.3b,c). Multiple-bond rupture events with various rupture lengths were observed in the FD curves recorded, on CsgA$^+$, using AFM tips conjugated with RGD, FN III, and FN (Figure 4.3d). In contrast, FD curves measurements, using the same tips, performed on CsgA$^-$ mutant showed a very low binding probability (less than 3%). The unbinding forces that originated from single-bond breakages with RGD, FN III, FN mostly fell in a force window between 45–60 pN (Figure 4.3e). This compares nicely with the forces observed for monomeric CsgA and implies that the interaction between RGD and CsgA drives bacterial adhesion when

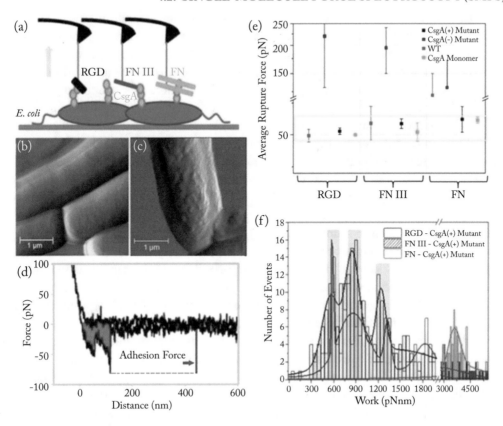

Figure 4.3: (a) AFM tips contained RGD, FN III, FN; surface-bound bacteria were WT, CsgA$^-$, and CsgA$^+$. (b) AFM deflection images of CsgA$^+$ mutant without curli expression. (c) CsgA$^+$ mutant after curli expression has been induced. (d) An example of FD curve recorded from RGD/CsgA$^+$ interactions. The red area indicates the adhesion work. (e) Single-molecular unbinding forces measured at a pulling speed of 500 nm/sec on each bacterial mutant using 2–3 different cantilevers and bacterial surfaces from 2–3 different batches. Force values from CsgA$^-$, WT, and CsgA were mainly collected within the yellow box indicated by a force range of 45–65 pN. The small number of detected forces on CsgA$^-$ (blue dots) showed irregular distributions indicative of multiple non-specific interactions that most likely arose from the production of flagella in this mutant. (f) Histogram of adhesion work for the dissociation of RGD, FN III, FN from the surface of CsgA$^+$ at a pulling velocity of 1000 nm/sec. Each distribution contains calculated adhesion work from 1000 force curves. The most probable adhesion work values from the maxima were fitted with multi-Gaussian distributions. Grey bars indicate work quanta. Single RGD required an adhesion work of 552 pN·nm, the second and third peak were at 840 pN · nm, and 1234 pN · nm. FN III showed maxima at 580, 856, and 1326. FN required much higher adhesion work and the most probable value was 3610 pN · nm [32].

curli fimbriae and fibronectin are involved. In addition, the non-equilibrium work required to detach the FN constructs from the curliated bacteria was calculated from the cumulative path integral of unbinding in FD curves (red area in FD Figure 4.3d). It includes contributions from deforming the bacterial membrane and from extending the curli fibers involved in molecular complexation, as well as the energy required for breaking all molecular connections. As shown in Figure 4.3f, histograms of the calculated adhesion work arising from the unbinding of RGD and FN III displayed characteristic maxima that were similarly distributed and consisted of three and four individual peaks of quantized nature, respectively. This implies that up to four tip-adorned molecules could access the bacterial membrane to contribute to the overall adhesion process. In contrast to RGD and FN III, the fully extended wild type FN showed a broad work distribution lacking resolution of individual bonds with the most probable value being about seven- to eight-fold the work quantum required for single RGD adhesion. The results of this study agreed well with the simple bond analysis model that suggests that the simultaneous breakage of N number of bonds occurs at a force less equal than N-times the force for breaking a single bond [37]. This decrease in work consumption per bond might indicate that the energy for membrane deformation was partially shared among the bonds, as expected from the parallel bond detachment observed. In addition, the overall work for detaching from the bacteria surface under mechanical force does not only contain contributions from the binding energy of the molecular bonds, but also from stretching the fibrous proteins and deforming the bacterial membrane. This study showed that curliated E. coli form quantized and multiple bonds of high tensile strength with fibronectin through specific multiple molecular connections via the RGD binding motif that lead to quasi-irreversible bacterial attachment. The insights provided by single molecule and microbial cell force spectroscopy measurements constitute the basis for unraveling novel mechanisms that govern bacteria-host cell interaction. This also offers exciting perspectives for controlling bacteria-host binding and thus opens new possibilities for alternative therapeutic strategies.

4.2.2   STRETCHING BACTERIAL PILI

Bacteria use multiple strategies to accomplish attachment and colonize surfaces [12, 38]. As shown in the previous example, curli expressed by many pathogenic isolates of *E. coli* and present on several *Salmonella* strains are mainly used to bind fibronectin receptors and do not generate motile forces. Other gram-negative bacteria use nanoscale protein fibers called type IV pili to generate a high motile force to swim close to host cells and induce bacterial internalization [12]. An important breakthrough in recent years, was the use of SMFS and SCFS to understand how these bacterial nanofibers respond to mechanical cues and how this response influences bacterial adhesion. Here we will focus on the well-known type IV gram-negative pili which mediate attachment of pathogens such as *Neisseria gonorrhoeae* and *P. aeruginosa* to host cells [12]. These pili are flexible rodlike filaments of 5–8 nm in diameter and 1–2 μm in length (Figure 4.4b), yet possess strong mechanical and adhesive properties [39]. Cooperative retraction of bundled pili

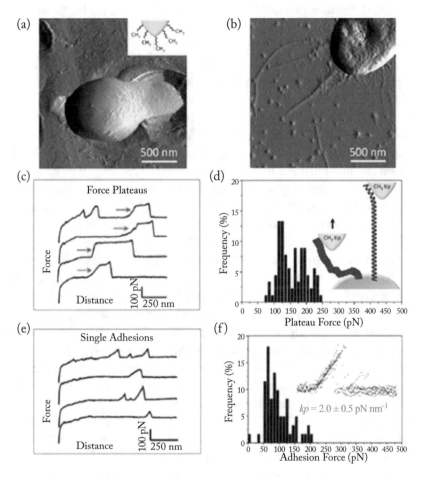

Figure 4.4: Quantifying the nanomechanics of type IV pili on living bacteria using chemical force microscopy. (a,b) AFM deflection images of wild-type *P. aeruginosa* bacteria recorded in buffer (a) and in air (b), showing that surface appendages can only be visualized in air. (c,d) Stretching individual pili yields constant force plateaus: typical plateau curves composed of a region at zero force followed by a progressive, nonlinear increase in the force (red arrows) to reach the constant force regime, and (d) histogram of the average plateau forces. As illustrated in the right panel, force plateau signatures are believed to result from force-induced conformational changes within the pili. (e,f) Stretched pili also show single linear force peaks, indicating that they behave as nanosprings: typical linear force peak signatures and histogram of maximum adhesion forces (f) as well as quantification of spring-like properties, estimation of pilus spring constant kp (inset). Superimposition of ten curves shows that spring-like properties are highly reproducible. Reprinted from ref. [12] with permission from Springer Nature. Copyright (2019).

can generate forces in the nanonewton range that could be critical for bacterial surface interactions [40].

Beaussart et al. [12] used both SMFS and SCFS to show that type IV pili strongly bind to hydrophobic surfaces in a time-dependent manner, while they weakly bind to hydrophilic surfaces. AFM tips were functionalized with hydrophobic groups (CH_3) and multiple FD curves were recorded in liquid medium on the poles of *P. aeruginosa* bacteria trapped in polymeric filter. FD curves recorded upon interaction with the hydrophobic tip, showed adhesion forces of 50–250 pN magnitude and 50–2000 nm rupture length. As shown if Figure 4.4c, some FD curves exhibited constant force plateaus, suggesting that the applied force caused conformational changes within individual pili filaments. Presumably, type IV pili resist mechanical force by transitioning into an extended quaternary structure that may expose adhesive residues [40]. This elongation model is consistent with the structure of gram-negative pili, known to result from non-covalent interactions between pilin subunits, and is also in line with earlier single-molecule studies on gram-negative pili [41–43]. However, the striking finding of this study, was the existence of single linear force peaks in some FD curves (Figure 4.4g,h), suggesting that type IV pili behave as nanosprings helping the bacteria to withstand physiological shear forces while being attached to its host. These findings emphasize the key role that type IV pili mechanics play in controlling bacterial attachment to biotic and abiotic surfaces.

In another pioneering study conducted by the same above researchers, gram-positive pili were found to have very different mechanical properties than gram-negative pili. Pili from the probiotic *Lactobacillus rhamnosus* GG (LGG) showed nanospring properties upon interaction with hydrophobic and mucin surfaces [44, 45]. Consistent with the covalent nature of gram-positive pili, this remarkable trait may help bacteria to withstand high mechanical forces in their natural environment. Pili stiffened at high forces, suggesting that external stress induces structural changes within individual fibers [45]. In addition, multiple adhesins (SpaC) distributed along the pilus length were shown to mediate zipper-like adhesion [44]. Adhesion to intestinal cells involved the extraction of nanotethers from the host membrane [45], a phenomenon expected to increase the lifetime of the interaction, thus favoring host colonization. Because human intestinal cells are largely covered with mucus, both nanosprings and nanotethers may contribute to LGG-host adhesion.

These studies demonstrate that gram-negative and gram-positive pili show fascinating mechanical behaviors that are expected to be important for cell adhesion. These experiments may have medical implications for the development of vaccines and anti-adhesion molecules capable of blocking the pilus activity.

4.3   SINGLE-CELL FORCE SPECTROSCOPY (SCFS)

4.3.1   SCFS PROTOCOL

SCFS consists of immobilizing a single living cell on an AFM cantilever and measuring the interaction forces between the cellular probe and a solid substrate or another cell. During the last two decades, various protocols have been developed to attach cells to AFM cantilevers [6, 9, 14, 18, 31]. The challenges are (i) to guarantee that the metabolic activity and natural surface architecture of the cells are preserved after cell immobilization and (ii) to carefully control the number of interacting cells. The lectin methodology, based on attaching cells on AFM tip-less cantilevers via specific receptor-ligand interactions [46], is often inappropriate for microbial cells because the cell-cantilever bond is too weak, leading to cell detachment during the measurements. Therefore, chemical fixation, glue, or drying procedures have been used by several research groups to attach cells to cantilevers [46–49]. For example, Ong and colleagues measured the hydrophobic forces between hard surfaces and bacteria chemically attached to cantilevers via glutaraldehyde [50]. The drying procedure is very simple and consists in immersing cantilevers in a concentrated cell suspension and allow them to dry for very short time [51, 52]. Adhesion forces of several microorganisms including *P. aeruginosa*, and *S. epidermidis* have been investigated using this methodology [51, 52]. Other methods based on electrostatic interactions using polymeric linkers such as poly-L-lysine and poly(ethyleneimine), were also used in number of SCFS studies [47, 53]. However, all of the above methods have some negative effects that may lead to cell surface denaturation, multiple cells are probed, or even cell death, raising questions about the biological relevance of the measurements.

Recently, new technology called fluidic force microscopy (FluidFM) offers exciting opportunities for non-invasive SCFS measurements [55–57]. It is based on microchanneled hollow cantilevers with nanosized apertures for manipulation of single living cells under physiological conditions. In microbiology, FluidFM has allowed researchers to quantify hydrophobic forces on Candida albicans cells [56]. Despite its strong potential for SCFS, FluidFM involves specific pieces of equipment that are not readily available on all microscopes.

During the last decade, a versatile protocol was developed for attaching single microbial cells to AFM cantilevers leading to reliable SCFS studies [9, 54]. As illustrated in Figure 4.5, a tip-less cantilever can be approached, using the standard AFM contact mode, to a drop of bioinspired polydopamine wet adhesive and just the apex of the cantilever is coated with the glue. Then, using the same mode a colloidal particle can be picked up and attached to the coated cantilever [58]. Using the same methodology, the sticky colloidal probe is used to pick up a single live cell from a cell suspension [58]. Fluorescence microscopy may be used to check that the cell is properly positioned and alive (Figure 4.5b, inset). Force-distance curves are then recorded on different areas of the target surface with the same cell probe, and the experiment is repeated with different surfaces and different probes to permit statistical analysis. This assay is non-destructive, enables single cell manipulation, and offers good control of the contact area, meaning that reproducible single cell analysis is possible (reference [54] gives a detailed protocol). This method was proved by the authors in analyzing the specific and non-specific forces of probiotic bacteria interacting with biotic or abiotic surfaces (Figure 4.5a). The key benefit for SCFS is that many cells are probed in a short timeframe, meaning that statistically relevant datasets can be obtained within a few hours. In the past few years SCFS has been instrumental in unraveling the forces in-

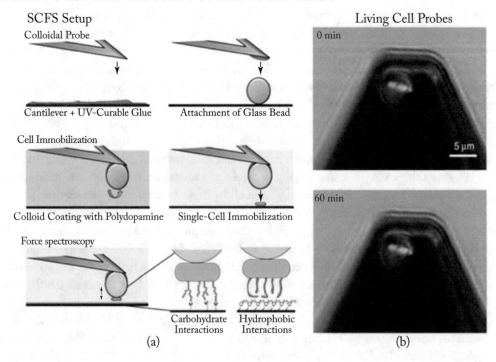

Figure 4.5: Principles of SCFS. (a) The SCFS protocol involves the use of colloidal probe cantilevers combined with bioinspired polydopamine polymers, and it consists of three steps: preparation of the colloidal probe, controlled attachment of single cells, and force-distance curve measurements. (b) Checking that the probed cells are still alive by using fluorescence stains: fluorescence images of bacterial cells labeled with the *Bac*light LIVE/DEAD stain, attached on polydopamine-coated cantilevers and imaged either immediately (0 min, top) or after 60 min of force measurements (bottom). Reprinted from ref. [54] with permission from Elsevier.

volved in microbe-microbe, microbe-host, and microbe-substrate interactions. Recently, SCFS was the subject of several breakthroughs in microbiological investigations of microbial adhesins, bacterial pili, and cell-cell associations.

4.3.2 FORCES DRIVING BACTERIA-HOST INVASION

In a recent elegant study, Prystopiuk et al. used several combinations of SCFS measurements (Figure 4.6) to determine the molecular forces driving the invasion of mammalian cells by *Staphylococcus aureus* [38]. It is known that this cellular invasion involves the interaction between the bacterial cell surface located fibronectin (Fn)-binding proteins (FnBPA and FnBPB) and the $\alpha 5\beta 1$ integrin at the mammalian cell surface. The measurement of molecular forces involved in this three component interaction revealed that the Fn bridge between FnBPA and

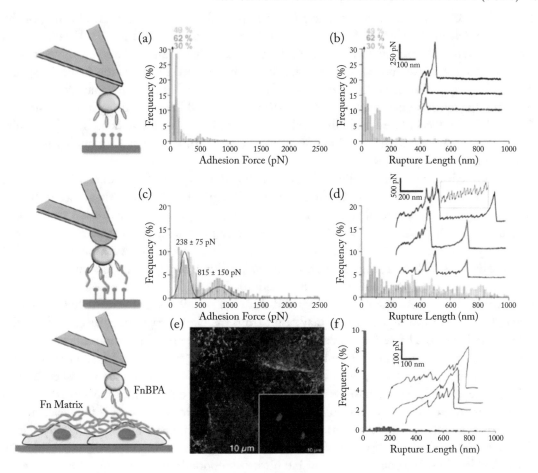

Figure 4.6: Mechanical stability of the Fn bridge between FnBPA and the $\alpha5\beta1$ integrin. (a,b) Adhesion force and rupture length histograms with representative force profiles obtained by recording force-distance curves in HEPES between S. aureus FnBPA(+) cells and $\alpha5\beta1$ integrins immobilized on solid substrates. Data from a total of 1244 curves from 3 different cells are shown. (c,d) Force data obtained between S. aureus FnBPA(+) cells pretreated with soluble Fn and integrin substrates (1171 curves from three cells). (e,f) Force data obtained between S. aureus FnBPA(−) cells pretreated with soluble Fn and integrin substrates (958 curves from three cells). (e) To study the interaction between FnBPA and Fn in the ECM, we measured the forces between FnBPA(+) bacterial probes and HUVEC monolayers spread on glass for 48 h. (f) Confocal microscopy image after staining with DAPI and anti-Fn antibody, documenting the production of large amounts of Fn (Fn is in green, nuclei in blue). The inset shows a control experiment in which primary anti-Fn antibody was missing. Reprinted with permission from ref. [38]. Copyright (2018) American Chemical Society.

the $\alpha5\beta1$ integrin is mechanically strong ($\sim$ 1500 pN), due to the cooperative loading of the multiple bonds of the tandem β-zipper formed between FnBPA and Fn [59]. Weaker bonds were observed with fibrillar Fn, suggesting that detection of β-zipper structures depends on the anchoring strength and conformational state of the protein. Fn forms mechanically strong bridges between FnBPAs on the *S. aureus* cell surface and purified integrins that are capable of withstanding much higher forces ($\sim$ 800 pN) than the classical Fn-$\alpha5\beta1$ integrin interaction ($\sim$ 100 pN). The force-induced unfolding and allosteric activation of FnIII domains. This results in the exposure of buried integrin-binding sites, which in turn engage in a strong, high-affinity interaction with integrins [38].

The high mechanical stability of the Fn bridge favors an invasion model in which Fn binding by FnBPA leads to the exposure of cryptic integrin-binding sites via allosteric activation, which in turn engage in a strong interaction with integrins. This activation mechanism emphasizes the importance of protein mechanobiology in regulating bacterial-host adhesion. The study also demonstrated that Fn-dependent adhesion between *S. aureus* and endothelial cells strengthens with time, suggesting that internalization occurs within a few minutes.

The interaction of S. aureus with host endothelial cells is also influenced by membrane deformation, which enhances the energy and duration of adhesion. Bacterial-host adhesion strengthens with time, an effect that may result from the clustering of integrins and from the internalization of the bacteria. The data from his study can be used for the identification of inhibitory compounds to treat infections involving intracellular pathogens. Most antibiotics are not effective in killing intracellular bacteria due to their poor penetration into the host cell membrane. An appealing approach to overcome this problem is to develop inhibitors capable of interfering with the Fn bridge. The study showed that the Fn-integrin bond is the weak side of the three-component complex, as the Fn-FnBPA tandem β-zipper is mechanically very stable. To prevent invasion the Fn bridge should be the main target via the integrin side rather than the FnBPA side. Thus, the combined use of inhibitory compounds and antibiotics could enhance antibiotic therapy.

4.4   MOLECULAR RECOGNITION MAPPING OF BACTERIA SURFACE

The molecular recognition mapping (MRM) mode was designed to generate single-molecule maps of specific types of molecules in a compositionally complex sample (such as bacteria surface), while simultaneously carrying out high-resolution topographic imaging [60]. The images are obtained in a dynamic force microscopy mode, and the target sites are recognized by sensor molecules covalently bound to the AFM tip [60]. However, there is less published work on the use of MRM compare to other force spectroscopy modes, due to the fact that SMFS can provides almost similar information and requires less preparations compare to MRM. Thus, most recent studies combine sets of measurements that involve mainly, SMFS and SCFS, modes to investigate single microbial cells [6–8, 25]. Nevertheless, mapping molecular recognition mea-

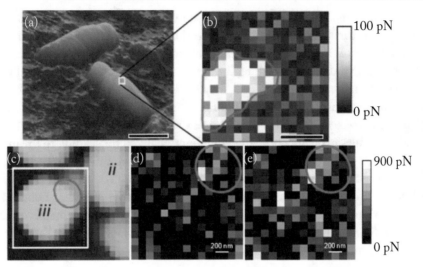

Figure 4.7: Mapping molecular recognition sites on living cells. Topographic image (left) showing two living mycobacteria on a polymer support and adhesion force map (right) recorded on a single cell with a heparin-modified tip. In localized regions, the map reveals adhesion events (clear pixels) owing to the presence of adhesion proteins referred to as heparin-binding haemagglutinin adhesin (HBHA). Notably, the adhesin distribution is not homogeneous, but apparently concentrated into nanodomains that may have an important role in mediating the attachment of mycobacteria to epithelial cells. Scale bars, 2 μm (left) and 100 nm (right). Adapted from [15]. (a) The circled area (red line) represents the contact zone between two bacteria (iii and ii in Figure 4.3a), where it is possible to observe an increase in the recognition events after 40 min (b and c). The same contact area in the elasticity maps (d and e) is circled by red lines. The envelope changes from soft to rigid values in the second scan (e) accompanying the raise in the recognition events for the domain as shown in b–c. Reprinted from ref. [62] with permission from Royal Society of Chemistry.

surements were among the first AFM experiments performed on living bacteria cells [61]. A pioneering study conducted by Dupres et al., who used heparin-modified AFM tip to map the distribution of single heparin-binding haemagglutinin adhesion (HBHA) on living Mycobacterium tuberculosis cells, found that the adhesins are not randomly distributed over the mycobacterial surface, but concentrated into nanodomains (Figure 4.7a,b) [15]. In a recent study, Arnal et al. probed the localization and distribution of the filamentous haemagglutinin (FHA) adhesin on the surface of living B. pertussis bacteria using an antibody-functionalized AFM tip [62]. As shown in Figure 4.7c–e, the authors reported that upon specific molecular recognition events take place (on the surface of a living cell) the FHA was reorganized and clustered with the time at the mapped area of the cell. As a consequence, the elasticity of the scanned area increased with time simultaneously. A critical issue when analyzing adhesion forces detected molecular recognition mapping AFM is to prove their specificity and to separate them from unspecific ones. Controls include blocking the specific interactions with antibodies or chemical compounds, and using mutant cells lacking the specific recognition sites. For direct comparison, fluorescently labeled target and mutant cells may be co-cultured, identified by fluorescence microscopy and simultaneously imaged with the functionalized tip. Tip contamination is another problem that needs to be addressed. With complex samples such as living cells, adsorption of loosely bound molecules may quickly change the functionalized tip, leading to the measurement of ill-defined tip-sample interactions. Therefore, before engaging functionalized tips, it is useful to characterize the sample with unmodified tips. Also, the applied force should be kept below 100 pN. Although adhesion force mapping provides a quantitative analysis of unbinding forces, it is limited by its time resolution. The time currently required to record a map is on the order of 2–15 min, depending on the acquisition parameters, which is much greater than the time scale at which dynamic processes usually occur in biology.

4.5   REFERENCES

[1] Xiao, J. and Dufrêne, Y. F. Optical and force nanoscopy in microbiology. *Nat. Microbiol.*, 1, 2016. 53, 54

[2] Cloud-Hansen, K. A. et al. Breaching the great wall: Peptidoglycan and microbial interactions. *Nat. Rev. Microbiol.*, 4:710–716, 2006. DOI: 10.1038/nrmicro1486.

[3] Kolter, R. and Greenberg, E. P. The superficial life of microbes. *Nature*, 441:300–302, 2006. DOI: 10.1038/441300a. 53

[4] Jamal, M. et al. Bacterial biofilm and associated infections-NC-ND license. *J. Chin. Med. Assoc.*, 81(1):7–11, http://creativecommons.org/licenses/by-nc-nd/4.0/, 2017. DOI: 10.1016/j.jcma.2017.07.012. 53

[5] Srivastava Atul Bhargava, S. Biofilms and human health. *Biotechnol. Lett.*, 38, 1960. DOI: 10.1007/s10529-015-1960-8. 53

[6] Dufrêne, Y. F., Martínez-Martín, D., Medalsy, I., Alsteens, D., and Müller, D. J. Multiparametric imaging of biological systems by force-distance curve-based AFM. *Nat. Methods*, 10:847–854, 2013. DOI: 10.1038/nmeth.2602. 53, 55, 56, 63, 66

[7] Dufrêne, Y. F. et al. Imaging modes of atomic force microscopy for application in molecular and cell biology. *Nat. Nanotechnol.*, 12:295–307, 2017. DOI: 10.1038/nnano.2017.45. 53, 54, 55, 56, 57, 58

[8] Alsteens, D., Müller, D. J., and Dufrêne, Y. F. Multiparametric atomic force microscopy imaging of biomolecular and cellular systems. *Acc. Chem. Res.*, 50:924–931, 2017. DOI: 10.1021/acs.accounts.6b00638. 53, 55, 56, 66

[9] Beaussart, A. et al. Quantifying the forces guiding microbial cell adhesion using single-cell force spectroscopy. *Nat. Protoc.*, 9:1049–1055, 2014. DOI: 10.1038/nprot.2014.066. 53, 54, 56, 63

[10] Hsieh, S. et al. Advances in cellular nanoscale force detection and manipulation. *Arab. J. Chem.*, 2015. DOI: 10.1016/j.arabjc.2015.08.011. 54

[11] Dufrêne, Y. F. Atomic force microscopy in microbiology: New structural and functional insights into the microbial cell surface. *MBio*, 5:1–14, 2014. DOI: 10.1128/mbio.01363-14. 54, 56, 58

[12] Beaussart, A. et al. Nanoscale adhesion forces of pseudomonas aeruginosa type IV pili. *ACS Nano*, 8:10723–10733, 2014. DOI: 10.1021/nn5044383. 60, 61, 62

[13] Müller, D. J. and Dufrêne, Y. F. Atomic force microscopy: A nanoscopic window on the cell surface. *Trends Cell Biol.*, 21:461–469, 2011. DOI: 10.1016/j.tcb.2011.04.008. 54

[14] Dufrêne, Y. F. Sticky microbes: Forces in microbial cell adhesion. *Trends Microbiol.*, 23:376–382, 2015. DOI: 10.1016/j.tim.2015.01.011. 63

[15] Dupres, V. et al. Nanoscale mapping and functional analysis of individual adhesins on living bacteria. *Nat. Meth.*, 2:515–520, 2005. DOI: 10.1038/nmeth769. 54, 56, 58, 67

[16] Alsteens, D. et al. Nanoscale imaging of microbial pathogens using atomic force microscopy. *WIREs Nanomed. Nanobiotech.*, 1:168–180, 2009. DOI: 10.1002/wnan.18. 58

[17] Wang, C. and Yadavalli, V. K. Investigating biomolecular recognition at the cell surface using atomic force microscopy. *Micron*, 60:5–17, 2014. DOI: 10.1016/j.micron.2014.01.002. 55

[18] Dufrêne, Y. F. Atomic force microscopy and chemical force microscopy of microbial cells. *Nat. Protoc.*, 3:1132–1138, 2008. DOI: 10.1038/nprot.2008.101. 54, 56, 63

[19] Dufrêne, Y. F. and Garcia-Parajo, M. F. Recent progress in cell surface nanoscopy: Light and force in the near-field. *Nano Today*, 7:390–403, 2012. DOI: 10.1016/j.nantod.2012.08.002. 54

[20] Butt, H.-J., Cappella, B., and Kappl, M. Force measurements with the atomic force microscope: Technique, interpretation and applications. *Surf. Sci. Rep.*, 59:1–152, 2005. DOI: 10.1016/j.surfrep.2005.08.003. 54, 55, 57

[21] Cappella, B. and Dietler, G. Force-distance curves by atomic force microscopy. *Surf. Sci. Rep.*, 34:1–3, 1999. DOI: 10.1016/s0167-5729(99)00003-5. 54, 57

[22] Radmacher, M., Cleveland, J. P., Fritz, M., Hansma, H. G., and Hansma, P. K. Mapping interaction forces with the atomic force microscope. *Biophys. J.*, 66:1–7, 2005. DOI: 10.1016/s0006-3495(94)81011-2. 55, 56

[23] Reese, R. A. and Xu, B. Single-molecule detection of proteins and toxins in food using atomic force microscopy. *Trends Food Sci. Technol.*, 83:277–284, 2018. DOI: 10.1016/j.tifs.2018.01.005. 56

[24] Müller, D. J., Helenius, J., Alsteens, D., and Dufrêne, Y. F. Force probing surfaces of living cells to molecular resolution. *Nat. Chem. Biol.*, 5:383–390, 2009. DOI: 10.1038/nchembio.181. 56

[25] Beaussart, A., Abellán-Flos, M., El-Kirat-Chatel, S., Vincent, S. P., and Dufrêne, Y. F. Force nanoscopy as a versatile platform for quantifying the activity of antiadhesion compounds targeting bacterial pathogens. *Nano Lett.*, 16:1299–1307, 2016. DOI: 10.1021/acs.nanolett.5b04689. 56, 58, 66

[26] Kumar, S. and Li, M. S. Biomolecules under mechanical force. *Phys. Rep.*, 486:1–74, 2010. DOI: 10.1016/j.physrep.2009.11.001. 57

[27] Bustamante, C., Macosko, J. C., and Wuite, G. J. L. Grabbing the cat by the tail: Manipulating molecules one by one. *Nat. Rev. Mol. Cell Biol.*, 1:130–136, 2000. DOI: 10.1038/35040072. 57

[28] Rief, M., Gautel, M., Oesterhelt, F., Fernandez, J. M., and Gaub, H. E. Reversible unfolding of individual titin immunoglobulin domains by AFM. *Science*, 276:1109–1112, 1997. DOI: 10.1126/science.276.5315.1109. 58

[29] Dufrêne, Y. F. Microbial nanoscopy: Breakthroughs, challenges, and opportunities. *ACS Nano*, 11:19–22, 2017. DOI: 10.1021/acsnano.6b08459. 58

[30] Müller, D. J. and Dufrêne, Y. F. Force nanoscopy of living cells. *Curr. Biol.*, 21:R212–R216, 2011. DOI: 10.1016/j.cub.2011.01.046. 58

[31] Alsteens, D., Van Dijck, P., Lipke, P. N., and Dufrêne, Y. F. Quantifying the forces driving cell-cell adhesion in a fungal pathogen. *Langmuir*, 29:13473–13480, 2013. DOI: 10.1021/la403237f. 58, 63

[32] Oh, Y. J. et al. Curli mediate bacterial adhesion to fibronectin via tensile multiple bonds. *Sci. Rep.*, 6:1–8, 2016. DOI: 10.1038/srep33909. 58, 59

[33] Gebbink, M. F. B. G., Claessen, D., Bouma, B., Dijkhuizen, L., and Wösten, H. A. B. Amyloids—A functional coat for microorganisms. *Nat. Rev. Microbiol.*, 3:333–341, 2005. DOI: 10.1038/nrmicro1127. 58

[34] Lipke, P. N. et al. Strengthening relationships: Amyloids create adhesion nanodomains in yeasts. *Trends Microbiol.*, 20:59–65, 2012. DOI: 10.1016/j.tim.2011.10.002.

[35] Gilchrist, K. B., Garcia, M. C., Sobonya, R., Lipke, P. N., and Klotz, S. A. New features of invasive candidiasis in humans: Amyloid formation by fungi and deposition of serum amyloid P component by the host. *J. Infect. Dis.*, 206:1473–1478, 2012. DOI: 10.1093/infdis/jis464.

[36] Gophna, U., Oelschlaeger, T. A., Hacker, J., and Ron, E. Z. Role of fibronectin in curli-mediated internalization. *FEMS Microbiol. Lett.*, 212(1):55–58, 2002. DOI: 10.1111/j.1574-6968.2002.tb11244.x. 58

[37] Williams, P. M. Analytical descriptions of dynamic force spectroscopy: Behaviour of multiple connections. *Anal. Chim. Acta*, 479:107–115, 2003. DOI: 10.1016/s0003-2670(02)01569-6. 60

[38] Prystopiuk, V. et al. Mechanical forces guiding staphylococcus aureus cellular invasion. *ACS Nano*, 12:3609–3622, 2018. DOI: 10.1021/acsnano.8b00716. 60, 64, 65, 66

[39] Touhami, A., Jericho, M. H., Boyd, J. M., and Beveridge, T. J. Nanoscale characterization and determination of adhesion forces of pseudomonas aeruginosa pili by using atomic force microscopy. *J. Bacteriol.*, 188:370–377, 2006. DOI: 10.1128/jb.188.2.370-377.2006. 60

[40] Biais, N., Ladoux, B., Higashi, D., So, M., and Sheetz, M. Cooperative retraction of bundled type IV pili enables nanonewton force generation. *PLoS Biol.*, 6:907–913, 2008. DOI: 10.1371/journal.pbio.0060087. 62

[41] Biais, N., Higashi, D. L., Brujić, J., So, M., and Sheetz, M. P. Force-dependent polymorphism in type IV pili reveals hidden epitopes. *Proc. Natl. Acad. Sci.*, 107:11358–11363, 2010. DOI: 10.1073/pnas.0911328107. 62

[42] Lugmaier, R. A., Schedin, S., Kühner, F., and Benoit, M. Dynamic restacking of escherichia coli P-pili. *Eur. Biophys. J.*, 37:111–120, 2008. DOI: 10.1007/s00249-007-0183-x.

[43] Miller, E., Garcia, T., Hultgren, S., and Oberhauser, A. F. The mechanical properties of E. coli type 1 pili measured by atomic force microscopy techniques. *Biophys. J.*, 91:3848–3856, 2006. DOI: 10.1529/biophysj.106.088989. 62

[44] Tripathi, P. et al. Adhesion and nanomechanics of pili from the probiotic lactobacillus rhamnosus GG. *ACS Nano*, 7:3685–3697, 2013. DOI: 10.1021/nn400705u. 62

[45] Sullan, R. M. A. et al. Single-cell force spectroscopy of pili-mediated adhesion. *Nanoscale*, 6:1134–1143, 2014. DOI: 10.1039/c3nr05462d. 62

[46] Helenius, J., Heisenberg, C. P., Gaub, H. E., and Muller, D. J. Single-cell force spectroscopy. *J. Cell Sci.*, 121:1785–1791, 2008. DOI: 10.1242/jcs.030999. 63

[47] Le, D. T. L., Guérardel, Y., Loubire, P., Mercier-Bonin, M., and Dague, E. Measuring kinetic dissociation/association constants between lactococcus lactis bacteria and mucins using living cell probes. *Biophys. J.*, 101:2843–2853, 2011. DOI: 10.1016/j.bpj.2011.10.034. 63

[48] Bowen, W. R., Lovitt, R. W., and Wright, C. J. Atomic force microscopy study of the adhesion of saccharomyces cerevisiae. *J. Colloid Interface Sci.*, 237:54–61, 2001. DOI: 10.1006/jcis.2001.7437.

[49] Bowen, W. R., Hilal, N., Lovitt, R. W., and Wright, C. J. Direct measurement of the force of adhesion of a single biological cell using an atomic force microscope. *Colloids Surf. A Physicochem. Eng. Asp.*, 136:231–234, 1998. DOI: 10.1016/s0927-7757(97)00243-4. 63

[50] Ong, Y. L., Razatos, A., Georgiou, G., and Sharma, M. M. Adhesion forces between E. coli bacteria and biomaterial surfaces. *Langmuir*, 15:2719–2725, 1999. DOI: 10.1021/la981104e. 63

[51] Emerson IV, R. J. et al. Microscale correlation between surface chemistry, texture, and the adhesive strength of staphylococcus epidermidis. *Langmuir*, 22:11311–11321, 2006. DOI: 10.1021/la061984u. 63

[52] Emerson, R. J. and Camesano, T. A. Nanoscale investigation of pathogenic microbial adhesion to a biomaterial. *Appl. Environ. Microbiol.*, 70:6012–6022, 2004. DOI: 10.1128/aem.70.10.6012-6022.2004. 63

[53] Ovchinnikova, E. S., Krom, B. P., Van Der Mei, H. C., and Busscher, H. J. Force microscopic and thermodynamic analysis of the adhesion between pseudomonas aeruginosa and candida albicans. *Soft Matter*, 8:6454–6461, 2012. DOI: 10.1039/c2sm25100k. 63

[54] Beaussart, A. et al. Single-cell force spectroscopy of probiotic bacteria. *Biophys. J.*, 104:1886–1892, 2013. DOI: 10.1016/j.bpj.2013.03.046. 63, 64

[55] Guillaume-Gentil, O. et al. Force-controlled manipulation of single cells: From AFM to FluidFM. *Trends Biotechnol.*, 32:381–388, 2014. DOI: 10.1016/j.tibtech.2014.04.008. 63

[56] Dörig, P. et al. Force-controlled spatial manipulation of viable mammalian cells and micro-organisms by means of FluidFM technology. *Appl. Phys. Lett.*, 97:023701, 2010. DOI: 10.1063/1.3462979. 63

[57] Meister, A. et al. FluidFM: Combining atomic force microscopy and nanofluidics in a universal liquid delivery system for single cell applications and beyond. *Nano Lett.*, 9:2501–2507, 2009. DOI: 10.1021/nl901384x. 63

[58] Lee, H., Lee, B. P., and Messersmith, P. B. A reversible wet/dry adhesive inspired by mussels and geckos. *Nature*, 448:338–341, 2007. DOI: 10.1038/nature05968. 63

[59] Sinha, B. and Herrmann, M. Mechanism and consequences of invasion of endothelial cells by staphylococcus aureus. *Thrombosis Haemostasis*, 94:266–277, 2005. DOI: 10.1160/th05-04-0235. 66

[60] Hinterdorfer, P. and Dufrêne, Y. F. Detection and localization of single molecular recognition events using atomic force microscopy. *Nat. Meth.*, 3:347–355, 2006. DOI: 10.1038/nmeth871. 66

[61] Ebner, A. et al. Localization of single avidin-biotin interactions using simultaneous topography and molecular recognition imaging. *ChemPhysChem*, 6:897–900, 2005. DOI: 10.1002/cphc.200400545. 67

[62] Arnal, L. et al. Localization of adhesins on the surface of a pathogenic bacterial envelope through atomic force microscopy. *Nanoscale*, 7:17563–17572, 2015. DOI: 10.1039/c5nr04644k. 67

CHAPTER 5

Bacteria Mechanics at the Nanoscale

OVERVIEW

Mechanical properties play a critical role in bacterial processes including cell growth, motility, division, and adaptation. For example, cell elasticity and its alterations were increasingly used during the last decades as a quantitative marker to describe the state and phenotype of cells. Historically, the bacteria cell mechanics were primarily attributed to the peptidoglycan which is known to stabilize the cell wall and provide structural integrity to the cell. In the current model of bacterial cell mechanics a diverse group of structural elements have been uncovered, indicating that the peptidoglycan is one element of a larger set of macromolecular materials that influence cell mechanics and vitality. To reach the desired level of comprehension, studies of cell mechanics and function must be paralleled by detailed investigations of specific molecular physical properties and interactions. Several techniques have been used to study the underlying mechanisms of cell mechanics, at the single cell level, such as optical tweezers, magnetic twisting cytometry, micropipette aspiration or optical stretcher technique [1–4]. Among all, AFM combines precise spatial resolution with high force sensitivity allowing the investigation of mechanical properties of living adherent cells in a unique fashion [3–5]. As detailed in the first chapter of this book, besides quantifying the inter- and intra-molecular interaction forces of biological systems, the AFM tip can be indented into and retracted from the sample to probe mechanical properties under almost physiological conditions. Such indentation-retraction experiments can provide insight into the deformation, elastic modulus, viscoelasticity, pressure, friction, adhesion, and energy dissipation of a biological sample [5]. These mechanical parameters can be mapped with a spatial resolution ranging from millimeters to sub-nanometers and at kinetic ranges from hours to milliseconds. We were among the first to study microbial cell wall mechanical properties by AFM. We showed that a daughter cell (bud) is ten times stiffer (6 MPa) than the surrounding cell wall formed on the parent yeast cell [4]. The stiffness increase was attributed to the accumulation of chitin which also plays an important structural role in the exoskeleton of arthropods and insects. It is worth to mention that AFM mechanical measurements are sensitive to experimental conditions, and it is preferable to compare data recorded on samples that have been prepared similarly. For example, dehydrating a sample can lead to a significant increase in cell stiffness, and the ionic strength of the media used in experiments can drastically alter the measured stiffness. In addition, in some cases noticeable discrepancies between experimental [6] and theoretical studies [7] have been reported that can be related both to the experimental conditions and the heterogeneity of the bacterial surface. For example, the large dispersion of young's modulus values for bacterial cells can simply reflects the heterogeneous composition of the cell envelope constituted in most bacteria by one or two layers of membrane, a rigid cell wall consisting of a network of peptidoglycan polymers and various appendices (pili, etc.). These components have different deformation values when loads are applied leading to a large dispersion in elastic modulus values [7].

This chapter highlights the most important studies that investigated the mechanical properties of several bacterial cell wall components using AFM-based nanomechanical measurements. We focus primarily on contributions that relates the physiology and pathology of bacteria to their nanomechanical properties.

5.1 BASICS OF AFM-SINGLE CELL NANOMECHANICAL MEASUREMENTS

Similar to all the previously described AFM force spectroscopy modes, AFM nanomechanical measurements require some specific preparations for both the sample and the cantilever. Although imaging and mechanical sensing by AFM might appear straightforward, several intricacies complicate the acquisition of quantitative data [3]. This section thus focuses on the key points that need to be taken into account to accomplish meaningful measurements.

5.1.1 SAMPLE PREPARATIONS AND REQUIREMENTS

The main challenges in performing AFM mechanical measurements on bacterial cells are: (i) the cells need to be hold in very stable way without altering their mechanical and physiological properties. Drying or chemically fixing of cells can lead to severe morphological, mechanical and functional artefacts [8–11]; (ii) cells should be in slow growing conditions to prevent cell division while performing measurements; and (iii) measurements must be performed at least on three different cells from the same sample, and three different samples using similar tip and cells growing in similar conditions. In addition, measurements should be carried out under almost physiological conditions required to maintain the native functional and morphological state of the cells. Such conditions mostly include full immersion of the sample in a buffer solution, an adjustable temperature and atmospheric control.

5.1.2 CHOOSING THE RIGHT CANTILEVER

Unlike other AFM force spectroscopy modes, it is trivial to choose the right cantilever for nanomechanical measurements. First, depending on the biological system under investigation and on the biological question, one can use probes with well-defined shapes and dimensions ranging from the micrometer to the nanometre scale (Figure 5.1a). If the indentation is so deep that the probe apex is entirely covered, or if the indentation depth is on the same scale as the

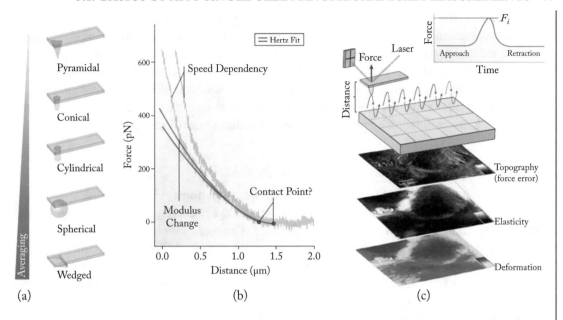

Figure 5.1: (a) Different probes can be used for the mechanical characterization of biological systems. The larger the probe contacting the sample, the more the measurement will average out over a larger sample area. (b) Typical challenges encountered when analyzing FD (or force-time) curves are defining the contact point, fitting the slope of the approach curve (different fits lead to different elastic moduli) and addressing changes in the apparent elastic modulus due to sample heterogeneity or inadequate data acquisition. A speed-dependent behavior indicates that the sample is viscoelastic. In this example, the grey curves are data acquired at different speeds. Blue and red curves are fits based on the Hertz model, which assume different contact points for the left-most grey curve. (c) FD-based atomic force microscopy can be used to contour the sample topography while measuring the elastic and inelastic deformation, viscoelasticity, energy dissipation, mechanical work, pressure and tension. For each pixel of the topography, at least one FD curve is recorded. (Fi, indentation force). Reprinted from ref. [3] by permission from Springer Nature. Copyright (2018).

roughness of the probe, it becomes notoriously difficult to estimate the sample deformation. By contrast, when the probe makes only a slight indentation (a few nanometers), contamination with macromolecules from the sample and buffer solution can alter its interaction with the sample. One solution to this problem is to routinely check for contamination by indenting reference samples while characterizing the biological system of interest [12]. The second critical requirement is to choose cantilevers that have spring constants very close to that of studied microbial cell. If the cantilever is much stiffer than the sample, the deflection becomes minimal, and the measurement insensitive, whereas cantilevers that are too soft do not sufficiently deform the sample, leading to difficulties in estimating the sample stiffness (Figure 5.1b). Several procedures for estimating the cantilever spring constant are available, including analyzing the thermal noise of the cantilever or pressing the cantilever against a reference cantilever [5, 13, 14]. Nevertheless, instrumental and experimental variabilities lead to considerable variations ($\sim$ 30%) between different laboratories in determining the spring constant of the same cantilever13. It is thus important to establish standardized procedures to determine cantilever spring constants and to check this calibration by probing reference samples or cantilevers [5].

5.1.3 HOW TO PROBE THE MECHANICAL PROPERTIES

As described in Figure 1.5 of Chapter 1, the basic way to perform AFM mechanical measurements is to indent the probe into the sample and to record FD curves, which reflect mechanical deformation and response of the sample under load. Force can also be plotted against time in force-time curves, which are particularly useful when the sample changes mechanical properties with time or viscoelastic properties need to be determined (Figure 5.1c) [15]. To extract the Young's modulus from the apparent stiffness measured by AFM, it is necessary to calculate the mechanical stress applied, which is the force per contact area of the probe and sample. However, the deeper a probe indents the sample, the more difficult it is to estimate how it interacts with and deforms the biological sample. Such estimations, which become notoriously difficult when using common pyramidal atomic force microscope probes, may be simplified by using cylindrical or spherical probes (Figure 5.1a) [3]. Indenting a sharp probe into a complex biological system enables the measurement of the mechanical properties only locally. The description of heterogeneous sample properties thus requires either multiple spatially discrete measurements or the use of larger probes to integrate properties over larger areas. For example, micrometer-sized spheres can be attached to an atomic force microscope cantilever [16]. Alternatively, the mechanical properties of entire cells, can be characterized by confining single cells between the parallel plates of a support and a wedged cantilever [9, 17]. To address the heterogeneity of biological systems, various AFM imaging modes have been introduced to map mechanical properties to morphology. The most common approach records at least one FD curve for every pixel of the AFM topography (Figure 5.1c). FD-based AFM can record several hundreds or thousands of force curves per topography, which makes the data analysis time consuming and requires automated procedures [18]. If the experiment has been conducted properly and a suitable model has been chosen for data analysis, topographs and multiparametric maps describing the mechanical properties of the sample are obtined [18, 19].

The mechanical properties of biological systems depend on the loading rate (the force increasing over time) at which they are measured. Because elastic, viscous and plastic components of complex systems respond differently to mechanical cues, the mechanical properties of cells and proteinaceous assemblies change nonlinearly with the loading rate (Figure 5.1b) [18–21]. Thus, it is meaningless to compare the mechanical properties of cells without specifying the loading rate. Varying the cantilever velocity can also enable differentiation between the possible underlying specific viscoelastic relations, such as linear or power-law rheology [22]. Additionally, complex materials respond differently to different mechanical stimuli (indentation, confinement, pressure, shear, friction, torsion, speed, or dynamic or nonlinear stimuli). Considering the anisotropic complexity of biological systems ranging from macromolecular complexes to living cells, tissues and organs, AFM experiments need to be designed carefully to apply well-defined mechanical cues and to characterize biomechanical properties over a wide range of loading rates. Another limitation is that AFM experiments mostly measure stress and strain as simple numbers, even though both are tensors describing how forces and deformation propagate in systems. The complex way forces deform structures such as macromolecules, cells, or tissues is difficult to describe without complementary experimental data and assumptions or extensive theoretical simulations [23].

5.1.4 MODELS TO EXTRACT MECHANICAL PROPERTIES

Although most commercial AFM software programs extract approximate mechanical parameters from force curves, the underlying models have several limitations. As detailed in the first chapter of this book, the most commonly used theoretical frameworks for approximating mechanical parameters from AFM measurements include the Hertz, Sneddon, Derjaguin–Müller–Toporov (DMT) and Johnson–Kendall–Roberts (JKR) models [24, 25]. Each model is applicable to different indenter geometries and sample properties. To assess mechanical properties, most AFMs measure the deformation of a sample in response to the force F applied by the indenting probe. Extracting the mechanical properties described by stress-strain curves from force curves requires a mechanical contact model, whereby the stress σ is approximated by the force per area and the deformation is approximated by the unitless strain ε. The Hertz model is the most frequently used to obtain mechanical parameters from AFM measurements. However, major assumptions are required such as, the probe is considered a perfect sphere perpendicularly indenting smooth plane surface. A second assumption is that the strain and elastic stress depend linearly on the Young's modulus $E (\sigma = E \cdot \varepsilon)$, which implies that the indentation must remain small compared with the dimensions of the sample and that the sample deformation must be fully reversible to ensure elasticity. However, complex biological structures such as living cells or tissues exhibit viscoelastic behavior, which manifests itself as a hysteresis between the approach and retraction FD curves. Thus, the stress-strain relationship must include the viscosity η such that $\sigma = \eta \cdot d\varepsilon/dt$. Another approximation of the Hertz model is that the contact area between the probe and sample is assumed to be much smaller than their dimensions. Finally, the Hertz model assumes that there are no other interactions, such as adhesion or friction, between the contacting surfaces. However, adhesion is often observed when cantilevers are pressed onto cells. Alternatively, the DMT or JKR models, which include adhesive effects, can be used.

Even if the above conditions for applying the Hertz model are met, accurate measurements require the careful control of experimental parameters. Extraction of the sample indentation from FD curves requires defining the point of contact (Figure 5.2b), which can be difficult to

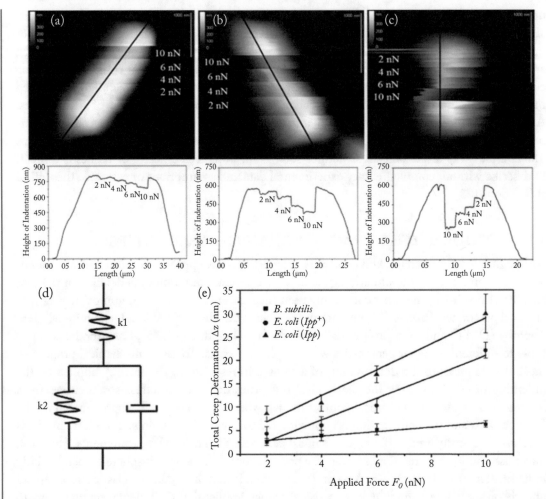

Figure 5.2: Representative examples of AFM topographic images of individual cells of (a) *B. subtilis 168*, (b) *E. coli* (lpp$^+$), and (c) *E. coli* (lpp) obtained in contact mode with an AFM pyramid-shaped tip by increasing the loading force F_0 in a stepwise fashion. The vertical white bars in each image correspond to 1 μm. The cross sections shown below each topographic image correspond to the black lines shown in the images along the length of the cell. The images and cross sections are intended to show the relative deformability of each bacterial strain, and no quantitative data were obtained from these images. (d) Schematic diagram of the standard solid model used to obtain the cell viscoelastic constants. (e) Total creep deformation Δz as a function of the loading force F_0 for the three bacteria strains measured using a colloidal AFM tip. The straight lines were calculated using the best-fit parameter values for each data set. The error bars correspond to the standard deviations of the average values obtained from the analysis of 15 creep curves for each loading force and cell type investigated [39].

determine. For example, as most living mammalian cells are compliant and have complex surface morphologies, a clear signature of the contact point can be missing from the FD curve, leading to an inaccuracy of a few tens of nanometers in the determination of the indentation depth. Typically, indentation depths of at least 400 nm are needed to avoid a dependence of the results on this inaccuracy.

5.2   MECHANICAL STRUCTURES IN BACTERIAL CELLS

Microbial cell surfaces differ substantially from the surfaces of mammalian cells; they are surrounded by thick, mechanically rigid cell walls, which play important roles in controlling cellular processes such as growth, division and adhesion. Almost without exception, peptidoglycan, the cross-linked polymeric meshwork that encapsulates bacterial cells, has historically been considered to be the canonical material in bacteria that imparts cell mechanical properties [26]. Additionally, the penicillin-binding proteins (PBPs) make up a class of proteins that affect the mechanical properties of the cell by directly altering the cross-linking and glycan strand length of the peptidoglycan. For example, insertion of new peptidoglycan is coordinated by the bacterial actin cytoskeleton homologue, MreB, which polymerize into filaments that rotate circumferentially around the long axis of cells [26]. The directed motion of MreB in these regions of the cell is correlated with peptidoglycan assembly and enables cells to maintain a rod shape. The identification of these diverse proteins indicates that cell stiffness is dependent on the function and coordination of numerous intracellular pathways and suggests a rich area for biochemical, biophysical, and cell biological studies.

Several other structural features and polymeric structures at the bacterial cell wall play critical roles maintaining mechanical properties of bacterial cells. In gram-negative bacteria, the outer membrane contains two types of lipids, lipopolysaccharide (LPS), and phospholipids, as well as other characteristic proteins [27]. Although the relationship between the stability of the LPS layer and its contribution to membrane permeability has been well established; however, there is a surprisingly small number of studies linking LPS to the mechanical properties of the cell [28]. Gram-negative bacteria cell surface can also be studded or covered by extracellular polymeric substances or some specific external structures (pili or fimbriae and flagella) that are essential for bacterial mechanics [28]. On the other hand, gram-positive bacteria do not have an outer membrane or LPS; however, they contain wall teichoic acids (WTA) and lipoteichoic acids (LTA) that are polysaccharides covalently attached to the peptidoglycan and inserted into the cytoplasmic membrane, respectively. WTAs account for $\sim 50\%$ of the weight of the cell wall, and play a role in membrane integrity [28]. Although little is presently known about the mechanical contribution of WTAs in gram-positive bacteria, the influence of this family of molecules on cell morphology and peptidoglycan thickness may affect the mechanical properties of the cell.

5.3 QUANTIFYING NANOMECHANICS OF CELL WALL COMPONENTS

Due to challenges in performing AFM nanomechanical measurements on living bacteria cells, in early days of atomic force microscopy the first studies focused mainly on quantitative measurement of the mechanical properties of isolated wall components, such as the proteinaceous sheath and murein sacculi. Thus, in pioneering study, Xu et al. [29] found remarkably very high Young's modulus values (20–40 GPa) for the sheath of the gram-negative *Methanospirillum hungatei* bacteria, suggesting that this cells can withstand an astonishingly high internal pressure of 400 atm. The same researchers investigated the relationship between thickness and elasticity of purified sacculi from *Escherichia coli K-12* and *Pseudomonas aeruginosa PAO1*. This prime study, showed that peptidoglycan sacculi are elastic structures that can expand most easily in the direction of the cell axis with increasing pressure [30]. The first AFM mechanical measurements on living cells were conducted by Amoldi et al. who measured the spring constant of the cell wall of *Magnetospirillum gryphiswaldense* bacteria to be, 42.10^{-3} N/m [31]. Abu-Lail et al. explored the elasticity and adhesion of the polymeric layer at the Pseudomonas putida surface in different potassium chloride concentrations. The elastic constant as well as the height of the brushed layer were found to change significantly with of the ionic strength of the solution [32]. Similarly, AFM-based indentation measurements were conducted by Gaboriaud et al. on *Shewanella putrefaciens* at two different pH values [33]. The study revealed an increase in height and a decrease in the stiffness of the bacterial envelope at high pH values, attributed to water exchange inside the polymeric fringe. In a similar study, the fibrillated *S. salivarius* bacteria surface was shown to be softer compare to nonfibrillated strain [34]. Lysostaphin treated Staphylococcus aureus bacteria were found less stiffer compare to untreated cells, indicating that digestion of peptidoglycan by the enzyme leads to the formation of osmotically fragile cells [35]. AFM has also proved useful for measuring local variations in the mechanical properties of live microorganisms. For example, Francius et al. demonstrated the crucial role played by cell wall polysaccharides in mechanical and adhesive properties of wild type *Lactobacillus rhamnosus* bacteria. The mutant CMPG5413 was found to be two times stiffer than the wild type LLG, presumably due to the absence of the surface exopolysaccharide layer from the mutant cells [36]. AFM indentation was also used to measure the elasticity changes of the surface of dormant *Bacillus anthracis* spores during germination process. The elasticity of the vegetative B. anthracis cells was nearly 15 times lower than that of the dormant spores (12.4 MPa vs. 197 MPa) [37]. This finding was consistent with the degradation of the multilayer surface of proteins and thick peptidoglycan coating the dormant Bacillus anthracis spores. Many other studies were conducted using AFM-based approaches to explore the morphological mapping of a wide variety of mechanical properties of bacteria cell wall components under physiologically relevant conditions. Here, we mentioned some of interesting early studies that were challenging to perform giving the limitations of the technique in the early days.

5.4   BACTERIAL VISCOELASTICITY AND TURGOR PRESSURE

In addition to their elastic behavior, bacterial cell walls also demonstrate a time dependent response to externally applied forces such that their mechanical properties are more properly described as viscoelastic [38]. It was demonstrated that both gram-positive and gram-negative bacteria exhibit viscoelastic responses to external compressive forces [27, 38–40]. In a typical AFM experiment, Dutcher and co-workers pioneered the investigation of viscoelastic properties of single bacteria cells under fluid conditions. The experiment consisted in recording the time dependent displacement (creep) of the AFM tip, due to the viscoelastic properties of the cell, while applying a constant compressive force to the cell under aqueous conditions [41]. The response of the *Pseudomonas aeruginosa PAO1*cells to the applied forces was well described by a three-element mechanical model consisted of an elastic spring, which describes an instantaneous elastic deformation (spring constant k_1), placed in series with a parallel combination of a spring and a dashpot, which describes a delayed elastic deformation (Figure 5.2). Treating the bacteria with glutaraldehyde, an agent that increases the covalent bonding of the cell surface, produced a significant increase in k_1 together with a significant decrease in the effective time constant for the creep deformation [41]. In a later study, the same authors performed a similar creep experiment in which they compared the local viscoelastic properties of individual gram-negative and gram-positive bacterial cells [39]. Large deformations of the cells were observed for all three strains, a gram-positive bacterium, *B. subtilis 168*, and two gram-negative cells, *E. coli* (lpp$^+$) and its Lpp mutant *E. coli* (lpp), as shown in Figure 5.2a–c. More pronounced morphological alterations were observed on the lpp mutant, defective in lipoprotein, indicating that the peptidoglycan-lipoprotein complex is required for maintaining the integrity of the cell wall structure. the gram-positive cell was the least deformable of the three strains (Figure 5.2a). Imaging the same cells at lower forces revealed that both gram-positive and gram-negative bacteria are able to support reversible nondestructive deformations depending on the composition and interconnection between the cell wall components. More importantly, the study demonstrated that the cells exhibit a viscoelastic solid-like behavior that is characterized by an instantaneous elastic deformation followed by a delayed elastic deformation. The instantaneous elastic response might be dominated by the properties of the peptidoglycan layer and the nature of its association with the membranes, whereas the delayed elastic response is more likely to arise from the liquid-like character of the cell membranes. This type of measurement can become a valuable technique for testing the effects of drugs that target different components of the cell envelope.

The intracellular hydrostatic pressure (turgor pressure), that results from the osmotic potential (concentration differential) across the membrane, plays trivial roles in cell growth and vitality. Several functions of bacterial cells are regulated by the turgor pressure, including bacterial signal transduction systems, bacterial periplasmic transport functions, and synthesis of porines [42]. It is therefore of paramount interest to measure the turgor pressure and to study the bacterial reactions to changes of osmolarity. Several techniques were used to estimate the bacterial turgor pressure under physiological conditions, including AFM indentation [43–48]. The estimated pressure values vary by more than an order of magnitude, from 104–3.105 Pa. While mechanical experiments, such as AFM indentation, are the most direct probes, separating the mechanical contributions of the wall and pressure has not been previously possible and thus these experiments may only provide an upper bound on the true turgor pressure. Shaevitz and co-workers, combined AFM and fluorescent microscopy to probe separately elasticity and turgor pressure of live *E. coli* cells (Figure 5.3). The study revealed that the cell wall in significantly stiffer in a living cell, young's modulus of 23 MPa and 49 MPa in the axial and circumferential directions respectively, compare to the purified sacculi [30]. However, the turgor pressure values, in living cells, were around 30 kPa, which are lower than previous chemical estimates of the pressure but similar to other mechanical measurements [43–46]. These data further indicate that the stress-stiffening affords a unique mechanical advantage to cells by preventing abrupt cell shape changes during changes in external pressure or osmolarity while maintaining a relatively compliant cell elasticity under normal conditions. In gram-positive bacteria, the turgor pressure is mainly supported by the thick peptidoglycan network which is the major cell-wall structural element [49]. Thus, changes in turgor pressure, due to changes in medium osmolarity, should result in changes in the nanoscale surface architecture and elastic properties of peptidoglycan. In order to investigate this assumption, Dover et al. used the multiparametric AFM-PeakForce tapping mode to probe the cell wall mechanical changes on live *Group B Streptococcus* (GBS) bacteria under different medium osmolarity (Figure 5.3) [47]. The PeakForce method enables higher sensitivity and minimal sample damage to soft biological samples due to a tight control over the force applied on the sample [50, 51]. The data showed a dynamic structural and mechanical response of peptidoglycan to elevation in cellular pressure. Under physiological salt concentration, the elastic modulus measurements suggested that the PG was in a less stressed state. Reducing the medium osmolarity increased the internal pressure and consequently a higher stress was put on the PG layers. Respectively, the cell wall became less elastic and the PG net expanded as the width of the circumferential bands doubled and the pore angles became on average straighter in water. A turgor-mediated increase in stiffness was also reported in *Escherichia coli* and yeast [43, 52]. The study revealed that a new nanoscale net-like arrangement of PG in live GBS, which stretches and stiffens to accommodate elevation in cellular pressure following an osmotic challenge.

5.5   PROBING MECHANICAL FORCES DURING CELL DIVISION

Cell division is a major process in the cell cycle of prokaryotes that occurs through significant morphological changes before full separation of the two daughter cells [26]. These morphological marks and their dynamics vary from one bacteria strain to another. For example, the last step in the division process of *E. coli* involves a gradual constriction in the middle of the elongated cell followed by structural remodeling of the new cell poles [53]. In contrast, in other bacteria

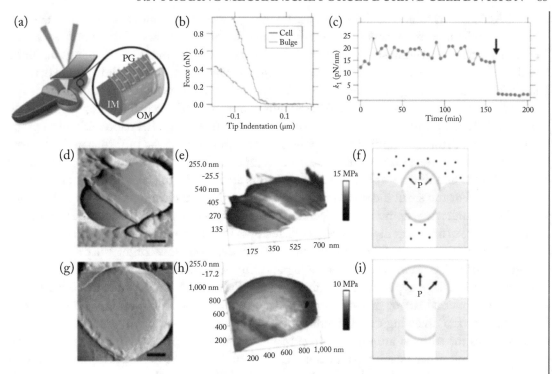

Figure 5.3: (a) Schematic cartoon illustrating the bulging *E. coli* and AFM stiffness measurement. The magnified region shows the details of the inner membrane (IM), peptidoglycan (PG) network and the outer membrane (OM). (b) Typical force-indentation traces obtained by indenting a cell and bulge. (f) Cell stiffness shows little variation before the bulging event (arrow) at which point it drops suddenly [43]. (d,g) Force-error images of single pore-trapped GBS in PBS and in ultrapure water, respectively (scale bar, 200 nm). (e,h) Directly correlated elastic moduli calculated from a 1 nN peak-force scan and rendered on a three-dimensional height representation of bacteria in PBS and ultrapure water, respectively. (f,i) Illustrations showing pore-trapped bacteria in high- or low-medium osmolarity, respectively. Elevation in turgor pressure (P) results in a higher force that swells the exposed surface [47].

species the two daughter cells remain connected until the septum is fully completed, with division scars appearing next to the new poles and no constriction [54–56]. Dividing a cell into two daughter cells requires mechanical work and enzymatic digestion of peptidoglycan, which is the main tensile stress-bearing component of the cell wall. Spatial and temporal control of mechanical and enzymatical effects is indispensable to maintain structural integrity of the cell wall [57].

AFM and its force modalities were used by many researchers to probe the structure and dynamics of cell wall components and live cells during bacterial growth and division processes [58–60]. In a recent pioneering study, Fantner and co-workers used fluorescence imaging combined with AFM quantitative nanomechanical mode (QNM) to study the growth and division dynamics of *Mycobacterium smegmatis*, similar to the human pathogen that causes tuberculosis [61]. The QNM mode provides simultaneous, high-resolution imaging and mechanical property mapping of live cell by recording force-distance curves at kilohertz rates [18]. The nanomechanical properties include deformation, dissipation, adhesion and modulus of the probed sample. Factors that contribute to the measured stiffness are the compound Young's modulus of the cell wall, the cell wall stress and the turgor pressure. These factors are thought to affect different parts of the force curve. For the small indentation depths used in QNM, the compound Young's modulus and the cell wall stress dominate the overall stiffness measurements [59]. The earliest morphological feature of a nascent division event detected by AFM imaging is the "pre-cleavage furrow" (PCF), a constriction of $\sim$ 50 nm width and 5–10 nm depth circumscribing the cell's short axis (Figure 5.4) [62]. Time-lapse AFM imaging and QNM measurements recorded at the site of the cell division showed progressive stiffening of the PCF over time, dependence of PCF stiffness on the internal turgor pressure, stress concentration at the PCF, and rapid (millisecond) physical cleavage of sibling cells (Figure 5.4). It is important to note, however, that the measured values for stiffness are very sensitive to factors such as cantilever tip geometry, sample topography and other imaging parameters. Care should therefore be taken in the interpretation to evaluate relative rather than absolute values within a time series. In summary, this study revealed that the PCF undergoes progressive stiffening before rapid cell cleavage, and that applied mechanical forces can induce premature cleavage, further strengthens the hypothesis that localized stress accumulation and mechanical fracture play a central role in mycobacterial cell division. The applied stress, however, is only one of the factors in fracture mechanics; the strength of the material is another important factor determining when and where fracture occurs. Thus, modulation of the ultimate tensile strength of the peptidoglycan by peptidoglycan synthesizing and hydrolytic enzymes is equally important to ensure that cell cleavage occurs at the appropriate time and place. Since molecular bonds under tensile stress require a lower activation energy for hydrolysis and are less likely to reform once broken [63], tensile stress around the PCF will locally accelerate the hydrolysis of peptidoglycan by the enzyme. As more chemical bonds are broken, the mechanical load on each remaining bond will increase, further facilitating hydrolysis. All together this data emphasis that localization of enzymes involved in cell wall biogenesis to subpolar growth zones determines where growth occurs [64], while the balance between cell wall stress and material strength determines when growth occurs. The overlapping and essential contributions of enzymatic hydrolytic activity and mechanical forces in mycobacterial cell division illustrate the general importance of studying molecular mechanisms and physical factors in conjunction rather than either in isolation.

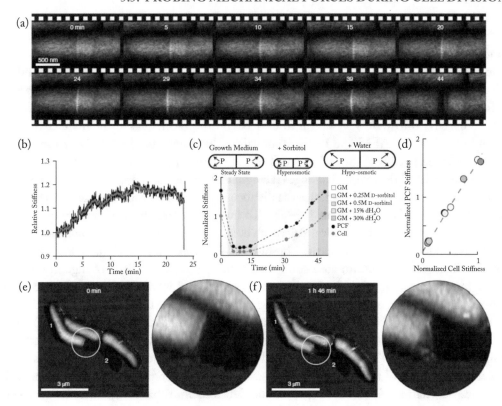

Figure 5.4: Turgor pressure drives cell cleavage through stress concentration at the PCF. (a) A time sequence of measured stiffness maps of the cell surface from the appearance of the PCF until cleavage. (b) Changes in the local stiffness at the PCF leading up to cell cleavage (y-axis, PCF stiffness normalized to bulk stiffness of cell sidewalls). After an initial gradual increase of the stiffness at the PCF, a small decrease occurs before cleavage (red line, average over 20 data points). (c) Measurement of apparent stiffness on the PCF (black lines) and on top of the cell bodies (blue lines, average of the two nascent sibling cells) as the turgor pressure is modulated by hyperosmotic (green shading) and hypo-osmotic (orange shading) media. The measured stiffness on the PCF and cell bodies is affected by the turgor pressure to a different extent. The measured stiffness values on the cell and on the PCF are normalized to the initial stiffness of the cell in normal growth medium. (d) When varying the medium osmolarity, the measured PCF stiffness is linearly dependent on the measured cell stiffness with a slope of ~ 1.5, suggesting a stress concentration of 50% on the PCF. The measured stiffness values on the cell and on the PCF are normalized to the initial stiffness of the cell in normal growth medium. The colors of the data points correspond to the shading in (c), (d). If one nascent sibling cell is deflated after cytokinesis but before cleavage (0 min), rapid cleavage is replaced by gradual shedding of the deflated sibling cell, indicating that rapid cleavage requires turgor pressure in both sibling cells. Reprinted from ref. [61] with permission from Springer Nature. Copyright (2019).

5.6   NANOMECHANICS OF BACTERIAL BIOFILM

A biofilm is a complex three-dimensional aggregation of microorganisms embedded in a protective matrix of exopolysaccharide (EPS) and adhering to a surface [65]. The structural and mechanical properties of the biofilm provide numerous advantages to bacteria, such as protection from antibiotics, disinfectants and dynamic environments [66]. AFM studies performed on single bacteria cells were very useful to understand the mechanics and dynamics of bacterial biofilms. However, AFM and its force modalities have been used to study mature biofilm structures since early days of the technique. The heterogeneity and gelatinous nature of mature biofilms were and still the major challenges facing AFM studies of these structures under aqueous conditions. Thus, early AFM studies usually focused on gaining topographical and morphological information of dried biofilms [67–69]. To investigate the effect of dehydration process on the overall structure of the biofilm, Auerbach et al. investigated the morphology and roughness of unsaturated fresh and desiccated biofilms of *P. putida* bacteria [70]. It was found that drying biofilms grown in unsaturated conditions (humid air) caused little change in morphology, roughness, or adhesion forces when compared with the moist biofilms, unlike the structural changes that occur when biofilms grown in fluid are dried for AFM analysis. Other studies were performed in liquid environments in the attempt to understand the more realistic properties, such as interaction and attachment to surfaces, of whole biofilms in aqueous conditions. For example, Ahimou et al. [71] developed an AFM-based technique for measuring, in situ, the cohesive energy levels of moist one-day biofilms. Cohesive strength is a primary factor which determines the relationship between growth and detachment of biofilms, where its quantification will aid in the understanding and modeling biofilm development. The biofilm was grown from an undefined mixed culture taken from activated sludge from a wastewater treatment plant. The volume of biofilm displaced, and the corresponding frictional energy dissipated were determined as a function of biofilm depth, after abrasion via a raster scanned tip under an elevated load. This resulted in the calculation of the cohesive energy level. Results showed that cohesive energy increased with biofilm depth, with four different biofilms showing the same behavior. Cohesive energy also increased when calcium was added to the reactor during biofilm growth. Arce et al. studied the pathogen *Nontypeable haemophilus* influenza through the use of AFM in an aqueous environment to image and characterize structural details, such as Hif-type pili, of the live bacteria at early stage of biofilm formation [72]. AFM force measurement has also been used to study bacterial biofilms. Volle et al. also performed force measurements on cells within a biofilm [67, 68]. Five bacterial strains, *E. coli ML35*, *E. coli ZK1056*, *P. putida*, *B. subtilis*, and *Micrococcus luteus*, were chosen for AFM analysis, which all formed simple biofilms on glass. The cells' spring constants and adhesion to the AFM tip were determined from a series of force curves in liquid. The spring constants and adhesion profiles of the five bacteria were distinctive, depending on the biology of the bacteria. The cellular spring constants varied between 0.16 and 0.41 N/m but it was noted that the gram-positive cells had larger spring constants than the gram-negative cells. Force measurements were also performed on both bacterial species of *E*.

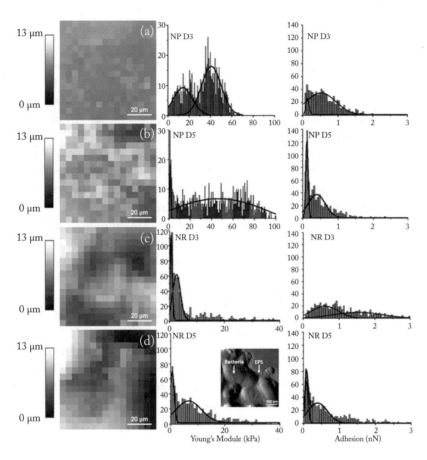

Figure 5.5: Examples of 80 × 80 μm force volume images (FVI) at a resolution of 16 × 16 pixels with their corresponding Young's modulus and adhesion histograms of (a) NP D3, (b) NP D5, (c) NR D3, and (d) NR D5. (inset is an AFM deflection image of the NR D5 biofilm) [73]. Individual bacteria can be identified due to their different morphological characteristics. FVI is a point-wise force-curved based imaging method, generating a force curve at each pixel from a chosen resolution. The advantage of AFM FVI is its capability to both image and obtain biofilm mechanical properties of a specimen simultaneously. Mechanical variation with respect to indentation location is shown by the elastic modulus grey scale. Indentation occurring on the bacteria surface show higher elastic modulus compared to those between bacteria. Young's modulus histograms of nutrient-poor and rich at 3 and 5 days of age are shown too. Increasing sucrose concentration was shown to significantly decrease Young's modulus at both stages of age. Binomial distributions are apparent in the histograms, indicating contact mechanics variations between the indenter and low- and high-bacteria density regions, respectively.

coli ZK1056 and *B. subtilis* in two different states: chemically fixed free-swimming planktonic cells and native biofilm cells without chemical fixation. Chemically fixed planktonic cells had different elasticity and adhesive profiles from the corresponding biofilm cells. This demonstrates that the physical properties of chemically fixed planktonic cells are significantly different from those of native biofilm cells.

In a recent study conducted by Pattem et al. focused on using optical coherence tomography (OCT) and AFM to understand the role sucrose concentration and age play in the morphological and mechanical properties of oral, microcosm biofilms, in vitro [73]. Distinct mesoscale features of biofilms such as regions of low and high extracellular polymeric substances (EPS) were identified through observations made by OCT (Figure 5.5). Mechanical analysis revealed increasing sucrose concentration decreased Young's modulus and increased cantilever adhesion, relative to the biofilm. Increasing age was found to decrease adhesion only (Figure 5.5). This was due to mechanical interactions between the indenter and the biofilm increasing as a function of increased EPS content, due to increasing sucrose. An expected, decrease in EPS cantilever contact decreased adhesion due to bacteria proliferation with biofilm age. The application OCT and AFM revealed new structure-property relationships in oral biofilms, unattainable if the techniques were used independently. It was found that increasing sucrose concentration increased biofilm deposition and significantly reduced Young's modulus. It also increased the adhesion of oral biofilms, while increasing age was shown to decrease adhesion only. This was associated with increasing EPS to bacteria ratios due to sucrose and decreasing this ratio due to aging. This analysis approach is not restricted to this specific type of biofilm, nor the in-vitro conditions in which future researchers wish to investigate. This approach can now be used in the future to elucidate the effect of potential removal strategies.

5.7   REFERENCES

[1] Designed Research. A S. J. A. contributed new reagents/analytic tools. 111, 2014. 75

[2] Tuson, H. H. et al. Measuring the stiffness of bacterial cells from growth rates in hydrogels of tunable elasticity. *Mol. Microbiol.*, 84:874–891, 2012. DOI: 10.1111/j.1365-2958.2012.08063.x.

[3] Krieg, M. et al. Atomic force microscopy-based mechanobiology. *Nat. Rev. Phys.*, 1:41–57. DOI: 10.1038/s42254-018-0001-7. 75, 76, 77, 78

[4] Touhami, A., Nysten, B., and Dufrêne, Y. F. Nanoscale mapping of the elasticity of microbial cells by atomic force microscopy, *Langmuir*, 19(11):4539–4543, 2003. DOI: 10.1021/la034136x. 75

[5] Schillers, H. et al. Standardized nanomechanical atomic force microscopy procedure (SNAP) for measuring soft and biological samples. *Sci. Rep.*, 7:1–9, 2017. DOI: 10.1038/s41598-017-05383-0. 75, 78

[6] Aguayo, S. and Bozec, L. Mechanics of bacterial cells and initial surface colonisation. *Exper. Med. Biol.*, 915:245–260, 2016. DOI: 10.1007/978-3-319-32189-9_15. 76

[7] Chen, Y., Norde, W., van der Mei, H. C., and Busscher, H. J. Bacterial cell surface deformation under external loading. *MBio 3*, 2012. DOI: 10.1128/mbio.00378-12. 76

[8] Sen, S., Subramanian, S., and Discher, D. E. Indentation and adhesive probing of a cell membrane with AFM: Theoretical model and experiments. *Biophys. J.*, 89:3203–3213, 2005. DOI: 10.1529/biophysj.105.063826. 76

[9] Fischer-Friedrich, E., Hyman, A. A., Jülicher, F., Müller, D. J., and Helenius, J. Quantification of surface tension and internal pressure generated by single mitotic cells. *Sci. Rep.*, 4, 2014. DOI: 10.1038/srep06213. 78

[10] Vorselen, D., Kooreman, E. S., Wuite, G. J. L., and Roos, W. H. Controlled tip wear on high roughness surfaces yields gradual broadening and rounding of cantilever tips. *Sci. Rep.*, 6:1–7, 2016. DOI: 10.1038/srep36972.

[11] Garcia, R. and Herruzo, E. T. The emergence of multifrequency force microscopy. *Nat. Nanotechn.*, 7:217–226, 2012. DOI: 10.1038/nnano.2012.38. 76

[12] Wischik, C. M. et al. Structural characterization of the core of the paired helical filament of Alzheimer disease. *Proc. Natl. Acad. Sci.*, 85:4884–4888, 1988. DOI: 10.1073/pnas.85.13.4884. 77

[13] te Riet, J. et al. Interlaboratory round robin on cantilever calibration for AFM force spectroscopy. *Ultramicroscopy*, 111:1659–1669, 2011. DOI: 10.1016/j.ultramic.2011.09.012. 78

[14] Calculation of thermal noise in atomic force microscopy—IOPscience. https://iopscience.iop.org/article/10.1088/0957--4484/6/1/001/meta, Accessed March 26, 2020. DOI: 10.1088/0957-4484/6/1/001. 78

[15] Fabry, B. et al. Scaling the microrheology of living cells. *Phys. Rev. Lett.*, 87:148102, 2001. DOI: 10.1103/physrevlett.87.148102. 78

[16] Mahaffy, R. E., Shih, C. K., MacKintosh, F. C., and Käs, J. Scanning probe-based frequency-dependent microrheology of polymer gels and biological cells. *Phys. Rev. Lett.*, 85:880–883, 2000. DOI: 10.1103/physrevlett.85.880. 78

[17] Stewart, M. P. et al. Wedged AFM-cantilevers for parallel plate cell mechanics. *Methods*, 60:186–194, 2013. DOI: 10.1016/j.ymeth.2013.02.015. 78

[18] Dufrêne, Y. F., Martínez-Martín, D., Medalsy, I., Alsteens, D., and Müller, D. J. Multiparametric imaging of biological systems by force-distance curve-based AFM. *Nat. Meth.*, 10:847–854, 2013. DOI: 10.1038/nmeth.2602. 78, 86

[19] Dufrêne, Y. F. et al. Imaging modes of atomic force microscopy for application in molecular and cell biology. *Nat. Nanotechn.*, 12:295–307, 2017. DOI: 10.1038/nnano.2017.45. 78

[20] Snijder, J., Ivanovska, I. L., Baclayon, M., Roos, W. H., and Wuite, G. J. L. Probing the impact of loading rate on the mechanical properties of viral nanoparticles. *Micron*, 43:1343–1350, 2012. DOI: 10.1016/j.micron.2012.04.011.

[21] Medalsy, I. D. and Müller, D. J. Nanomechanical properties of proteins and membranes depend on loading rate and electrostatic interactions. *ACS Nano*, 7:2642–2650, 2013. DOI: 10.1021/nn400015z. 78

[22] Efremov, Y. M., Wang, W. H., Hardy, S. D., Geahlen, R. L., and Raman, A. Measuring nanoscale viscoelastic parameters of cells directly from AFM force-displacement curves. *Sci. Rep.*, 7:1541, 2017. DOI: 10.1038/s41598-017-01784-3. 79

[23] Schoeler, C. et al. Mapping mechanical force propagation through biomolecular complexes. *Nano Lett.*, 15:7370–7376, 2015. DOI: 10.1021/acs.nanolett.5b02727. 79

[24] Sneddon, I. N. The relation between load and penetration in the axisymmetric boussinesq problem for a punch of arbitrary profile. *Int. J. Eng. Sci.*, 3:47–57, 1965. DOI: 10.1016/0020-7225(65)90019-4. 79

[25] Surface energy and the contact of elastic solids. *Proc. R. Soc. London. A. Math. Phys. Sci.*, 324:301–313, 1971. DOI: 10.1098/rspa.1971.0141. 79

[26] Cabeen, M. T. and Jacobs-Wagner, C. Bacterial cell shape, *Nat. Rev. Microbiol.*, 3:601–610, 2015. DOI: 10.1038/nrmicro1205. 81, 84

[27] Thwaites, J. J. and Mendelson, N. H. Mechanical behaviour of bacterial cell walls. *Adv. Microb. Physiol.*, 32:173–222, 1991. DOI: 10.1016/s0065-2911(08)60008-9. 81, 83

[28] Kline, K. A., Dodson, K. W., Caparon, M. G., and Hultgren, S. J. A tale of two pili: Assembly and function of pili in bacteria. *Trends Microbiol.*, 18:224-232, 2010. DOI: 10.1016/j.tim.2010.03.002. 81

[29] Xu, W. et al. Modeling and measuring the elastic properties of an archaeal surface, the sheath of Methanospirillum hungatei, and the implication for methane production. *J. Bacteriol.*, 178:3106–3112, 1996. DOI: 10.1128/jb.178.11.3106-3112.1996. 82

[30] Yao, X., Jericho, M., Pink, D., and Beveridge, T. Thickness and elasticity of gram-negative murein sacculi measured by atomic force microscopy. *J. Bacteriol.*, 181:6865–75, 1999. DOI: 10.1128/jb.181.22.6865-6875.1999. 82, 84

[31] Arnoldi, M., Kacher, C. M., Bäuerlein, E., Radmacher, M., and Fritz, M. Elastic properties of the cell wall of magnetospirillum gryphiswaldense investigated by atomic forcemicroscopy. *Appl. Phys. A Mater. Sci. Process.*, 66:613–617, 1998. DOI: 10.1007/s003390051210. 82

[32] Abu-Lail, N. I. and Camesano, T. A. Elasticity of pseudomonas putida KT2442 surface polymers probed with single-molecule force microscopy. *Langmuir*, 18:4071–4081, 2002. DOI: 10.1021/la015695b. 82

[33] Gaboriaud, F., Bailet, S., Dague, E., and Jorand, F. Surface structure and nanomechanical properties of shewanella putrefaciens bacteria at two pH values (4 and 10) determined by atomic force microscopy. *J. Bacteriol.*, 187:3864–3868, 2005. DOI: 10.1128/jb.187.11.3864-3868.2005. 82

[34] van der Mei, H. C. et al. Direct probing by atomic force microscopy of the cell surface softness of a fibrillated and nonfibrillated oral streptococcal strain. *Biophys. J.*, 78:2668–2674, 2000. DOI: 10.1016/s0006-3495(00)76810-x. 82

[35] Francius, G., Domenech, O., Mingeot-Leclercq, M. P., and Dufrêne, Y. F. Direct observation of staphylococcus aureus cell wall digestion by lysostaphin. *J. Bacteriol.*, 190:7904–7909, 2008. DOI: 10.1128/jb.01116-08. 82

[36] Francius, G. et al. Conformational analysis of single polysaccharide molecules on live. *Am. Chem. Soc. NANO*, 2:1921–1929, 2008. 82

[37] Pinzón-Arango, P. A., Nagarajan, R., and Camesano, T. A. Effects of L-alanine and inosine germinants on the elasticity of bacillus anthracis spores. *Langmuir*, 26:6535–6541, 2010. DOI: 10.1021/la904071y. 82

[38] Vadillo-Rodríguez, V. and Dutcher, J. R. Viscoelasticity of the bacterial cell envelope. *Soft Matter*, 7:4101–4110, 2011. DOI: 10.1039/c0sm01054e. 83

[39] Vadillo-Rodriguez, V., Schooling, S. R., and Dutcher, J. R. In situ characterization of differences in the viscoelastic response of individual gram-negative and gram-positive bacterial cells. *J. Bacteriol.*, 191:5518–5525, 2009. DOI: 10.1128/jb.00528-09. 80, 83

[40] Vadillo-Rodriguez, V. and Dutcher, J. R. Dynamic viscoelastic behavior of individual gram-negative bacterial cells. *Soft Matter*, 5:5012–5019, 2009. DOI: 10.1039/b912227c. 83

[41] Vadillo-Rodriguez, V., Beveridge, T. J., and Dutcher, J. R. Surface viscoelasticity of individual gram-negative bacterial cells measured using atomic force microscopy. *J. Bacteriol.*, 190:4225–4232, 2008. DOI: 10.1128/jb.00132-08. 83

[42] Rojas, E. R. and Huang, K. C. Regulation of microbial growth by turgor pressure. *Curr. Opin. Microbiol.*, 42:62–70, 2018. DOI: 10.1016/j.mib.2017.10.015. 83

[43] Deng, Y., Sun, M., and Shaevitz, J. W. Direct measurement of cell wall stress stiffening and turgor pressure in live bacterial cells. *Phys. Rev. Lett.*, 107:7–10, 2011. DOI: 10.1103/physrevlett.107.158101. 84, 85

[44] Yao, X. et al. Atomic force microscopy and theoretical considerations of surface properties and turgor pressures of bacteria. *Colloids Surf. B Biointerf.*, 23:213–230, 2002. DOI: 10.1016/s0927-7765(01)00249-1.

[45] Cayley, D. S., Guttman, H. J., and Record, M. T. Biophysical characterization of changes in amounts and activity of escherichia coli cell and compartment water and turgor pressure in response to osmotic stress. *Biophys. J.*, 78:1748–1764, 2000. DOI: 10.1016/s0006-3495(00)76726-9.

[46] Holland, D. P. and Walsby, A. E. Digital recordings of gas-vesicle collapse used to measure turgor pressure and cell-water relations of cyanobacterial cells. *J. Microbiol. Meth.*, 77:214–224, 2009. DOI: 10.1016/j.mimet.2009.02.005. 84

[47] Saar Dover, R., Bitler, A., Shimoni, E., Trieu-Cuot, P., and Shai, Y. Multiparametric AFM reveals turgor-responsive net-like peptidoglycan architecture in live streptococci. *Nat. Commun.*, 6, 2015. DOI: 10.1038/ncomms8193. 84, 85

[48] Arnoldi, M. et al. Bacterial turgor pressure can be measured by atomic force microscopy. *Phys. Rev. E—Stat. Physics, Plasmas, Fluids, Relat. Interdiscip. Top.*, 62:1034–1044, 2000. DOI: 10.1103/physreve.62.1034. 84

[49] Vollmer, W. and Seligman, S. J. Architecture of peptidoglycan: More data and more models. *Trends Microbiol.*, 18:59–66, 2010. DOI: 10.1016/j.tim.2009.12.004. 84

[50] Mapping, M. P., Pittenger, B., Erina, N., and Su, C. Application note # 128 quantitative mechanical property mapping at the nanoscale with PeakForce QNM. im:12, 2012. 84

[51] Alsteens, D., Trabelsi, H., Soumillion, P., and Dufrêne, Y. F. Multiparametric atomic force microscopy imaging of single bacteriophages extruding from living bacteria. *Nat. Commun.*, 4:2926, 2013. DOI: 10.1038/ncomms3926. 84

[52] Arfsten, J., Leupold, S., Bradtmöller, C., Kampen, I., and Kwade, A. Atomic force microscopy studies on the nanomechanical properties of saccharomyces cerevisiae. *Colloids Surf. B Biointerf.*, 79:284–290, 2010. DOI: 10.1016/j.colsurfb.2010.04.011. 84

[53] Gray, A. N. et al. Coordination of peptidoglycan synthesis and outer membrane constriction during escherichia coli cell division. *eLife*, 1–29, 2015. DOI: 10.7554/elife.07118.026. 84

[54] Dahl, J. L. Electron microscopy analysis of mycobacterium tuberculosis cell division. *FEMS Microbiol. Lett.*, 240:15–20, 2004. DOI: 10.1016/j.femsle.2004.09.004. 85

[55] Vijay, S., Anand, D., and Ajitkumar, P. Unveiling unusual features of formation of septal partition and constriction in mycobacteria-an ultrastructural study. *J. Bacteriol.*, 194:702–707, 2012. DOI: 10.1128/jb.06184-11.

[56] Monteiro, J. M. et al. Cell shape dynamics during the staphylococcal cell cycle. *Nat. Commun.*, 6:1–12, 2015. DOI: 10.1038/ncomms9055. 85

[57] Chao, M. C. et al. Protein complexes and proteolytic activation of the cell wall hydrolase RipA regulate septal resolution in mycobacteria. *PLoS Pathog.*, 9, 2013. DOI: 10.1371/journal.ppat.1003197. 85

[58] Touhami, A., Jericho, M. H., and Beveridge, T. J. Atomic force microscopy of cell growth and division in staphylococcus aureus. *J. Bacteriol.*, 186:3286–3295, 2004. DOI: 10.1128/jb.186.11.3286-3295.2004. 86

[59] Bailey, R. G. et al. The interplay between cell wall mechanical properties and the cell cycle in staphylococcus aureus. *Biophys. J.*, 107:2538–2545, 2014. DOI: 10.1016/j.bpj.2014.10.036. 86

[60] Turner, R. D., Hurd, A. F., Cadby, A., Hobbs, J. K., and Foster, S. J. Article cell wall elongation mode in gram-negative bacteria is determined by peptidoglycan architecture. *Nat. Commun.*, 4, 2013. DOI: 10.1038/ncomms2503. 86

[61] Odermatt, P. D. et al. Overlapping and essential roles for molecular and mechanical mechanisms in mycobacterial cell division. *Nat. Phys.*, 16:57–62, 2020. DOI: 10.1038/s41567-019-0679-1. 86, 87

[62] Eskandarian, H. A. et al. Division site selection linked to inherited cell surface wave troughs in mycobacteria. *Nat. Microbiol.*, 2:17094, 2017. DOI: 10.1038/nmicrobiol.2017.94. 86

[63] Koch, A. L. Biophysics of bacterial walls viewed as stress-bearing fabric. *Microbiol. Rev.*, 52:337–353, 1988. DOI: 10.1128/mmbr.52.3.337-353.1988. 86

[64] Meniche, X. et al. Subpolar addition of new cell wall is directed by DivIVA in mycobacteria. *Proc. Natl. Acad. Sci.*, 111, 2014. DOI: 10.1073/pnas.1402158111. 86

[65] Even, C. et al. Recent advances in studying single bacteria and biofilm mechanics. *Colloid Interf. Sci.*, 247:573–588, 2017. DOI: 10.1016/j.cis.2017.07.026. 88

[66] James, S. A., Powell, L. C., and Wright, C. J. Atomic force microscopy of biofilms—imaging, interactions, and mechanics, in *Microbial Biofilms: Importance and Applications*, 2016. DOI: 10.5772/63312. 88

[67] Volle, C. B., Ferguson, M. A., Aidala, K. E., Spain, E. M., and Núñez, M. E. Spring constants and adhesive properties of native bacterial biofilm cells measured by atomic force microscopy. *Colloids Surf. B Biointerf.*, 67:32–40, 2008. DOI: 10.1016/j.colsurfb.2008.07.021. 88

[68] Volle, C. B., Ferguson, M. A., Aidala, K. E., Spain, E. M., and Núñez, M. E. Quantitative changes in the elasticity and adhesive properties of escherichia coli ZK1056 prey cells during predation by Bdello vibrio bacteriovorus 109J. *Langmuir*, 24:8102–8110, 2008. DOI: 10.1021/la8009354. 88

[69] Méndez-Vilas, A. et al. Surface characterisation of two strains of staphylococcus epidermidis with different slime-production by AFM. *Appl. Surf. Sci.*, 238:18–23, Elsevier, 2004. DOI: 10.1016/j.apsusc.2004.05.183. 88

[70] Auerbach, I. D., Sorensen, C., Hansma, H. G., and Holden, P. A. Physical morphology and surface properties of unsaturated Pseudomonas putida biofilms. *J. Bacteriol.*, 182:3809–3815, 2000. DOI: 10.1128/jb.182.13.3809-3815.2000. 88

[71] Ahimou, F., Semmens, M. J., Novak, P. J., and Haugstad, G. Biofilm cohesiveness measurement using a novel atomic force microscopy methodology. *Appl. Environ. Microbiol.*, 73:2897–2904, 2007. DOI: 10.1128/aem.02388-06. 88

[72] Arce, F. T. et al. Nanoscale structural and mechanical properties of nontypeable haemophilus influenzae biofilms. *J. Bacteriol.*, 191:2512–2520, 2009. DOI: 10.1128/jb.01596-08. 88

[73] Pattem, J. et al. A multi-scale biophysical approach to develop structure-property relationships in oral biofilms. *Sci. Rep.*, 8:5691, 2018. DOI: 10.1038/s41598-018-23798-1. 89, 90

Author's Biography

AHMED TOUHAMI

Dr. Ahmed Touhami is an Associate Professor at the Department of Physics & Astronomy at the University of Texas Rio Grande Valley (UTRGV). He completed his undergraduate, Master's, and Ph.D. degrees in Physics at the Pierre & Marie Curie University (Paris-France). After joining Yves Dufrêne's lab at the Catholic University of Louvain-La-Neuve (Belgium) as a postdoctoral fellow, he moved to Dalhousie University (Canada) and then The University of Guelph (Canada), as an associate-research Fellow. Trained as a physicist and having worked in the fields of single molecule biophysics, biological physics, and nanoscience, he has considerable expertise in single molecule techniques such as Atomic Force Microscopy, Fluorescence Microscopy, and Optical Tweezers. He developed a new method to probe bacterial surface structures and dynamics in real time and under physiological conditions. He has published over 80 research papers, including book chapters, all based on single molecule biophysics and nanoscience studies. Dr. Touhami is a Member of Biophysical Society and the American Physical Society.

Printed in the United States
by Baker & Taylor Publisher Services